T0142907

Progress in Inflammation Research

Volume 87

The last few years have seen a revolution in our understanding of how blood and tissue cells interact and of the intracellular mechanisms controlling their activation. This has not only provided multiple targets for potential anti-inflammatory and immunomodulatory therapy, but has also revealed the underlying inflammatory pathology of many diseases.

This series provides up-to-date information on the latest developments in the pathology, mechanisms and therapy of inflammatory disease. Areas covered include: vascular responses, skin inflammation, pain, neuroinflammation, arthritis cartilage and bone, airways inflammation and asthma, allergy, cytokines and inflammatory mediators, cell signalling, and recent advances in drug therapy.

Each volume is edited by acknowledged experts providing succinct overviews on specific topics intended to inform and explain. The series is of interest to academic and industrial biomedical researchers, drug development personnel and rheumatologists, allergists, pathologists, dermatologists and other clinicians requiring regular scientific updates.

More information about this series at http://www.springer.com/series/4983

Carlo Rossetti • Francesco Peri
Editors

The Role of Toll-Like Receptor 4 in Infectious and Non Infectious Inflammation

Editors
Carlo Rossetti
Department of Medicine and Surgery
University of Insubria
Varese, Italy

Francesco Peri
Department of Biotechnology and Biosciences
University of Milano Bicocca
Milan, Italy

ISSN 1422-7746 ISSN 2296-4525 (electronic)
Progress in Inflammation Research
ISBN 978-3-030-56321-9 ISBN 978-3-030-56319-6 (eBook)
https://doi.org/10.1007/978-3-030-56319-6

This Springer imprint is published by the registered company Springer Nature Switzerland AG
The registered company address is: Gewerbestrasse 11, 6330 Cham, Switzerland

Preface

Toll-like receptors (TLRs) are essential regulators of innate and adaptive immune responses and their complexity continues to intrigue researchers, including the authors of this book. Because of their importance, when they are mutated and not functioning as they should, autoimmune, inflammatory, and infectious diseases can develop.

The aim of this book is to present an update on the role of TLR4, the most studied TLR, in inflammatory and infectious diseases with special focus on central nervous system (CNS) pathologies.

We also give an outlook on what emerged in the past years on the molecular aspects of extracellular TLR4 activation and intracellular signaling, its regulation by miRNA, and crosstalk with other metabolic pathways.

To this end, a group of internationally recognized experts has kindly accepted to present recent results on TLR4 function and role in health and disease.

TLR4 was the first TLR identified by Medzhitov and coworkers in 1997 and was then characterized by Beutler and coworkers as a pattern recognition receptor. TLR4 has an exquisite ligand selectivity that has remained largely unchanged through the course of evolution allowing for an immediate and sensitive response to gram-negative bacterial lipopolysaccharide (LPS).

The understanding of molecular features of extracellular TLR4 activation has contributed to unravel the physiological and pathological role of TLR4. We have now information on the structural biology of the molecular actors of LPS transfer: LBP and CD14 proteins, the M-shaped dimeric $(TLR4\text{-}MD2\text{-}LPS)_2$ complex, and of the intracellular signaling proteins, belonging to the so-called MyD88-dependent and MyD88-independent pathways. The LPS-binding protein CD14 not only takes part in LPS extraction from aggregates in solution and shuttling to TLR4/MD-2/LPS dimer, but it also exerts autonomous functions by regulating endocytic processes or by activating dedicated signaling pathways, as critically reviewed by M. Di Gioia and I. Zanoni.

Starting from the knowledge of the supramolecular interactions among LPS and LBP, CD14 and MD-2, the structure–activity relationship of TLR4 ligands, especially of lipid A variants, has been extensively studied. These studies allow the structure-based rational design of synthetic or semisynthetic lipid A variants as vaccine adjuvants as reviewed by A. Shimoyama and K. Fukase.

Natural TLR4 ligands, LPS and LOS, and their synthetic variants are amphiphilic molecules that aggregate in solution. A. B. Schromm and K. Brandenburg discuss from a biophysical perspective the most recent achievements in the study of the role of aggregates in the biological activity of TLR4 ligands, with special focus on the very recent findings on LPS interaction with intracellular caspases and subsequent induction of the non-canonical inflammasome.

To complete the complex and fascinating view of TLR4 signaling at a molecular level, N. Kuzmich dissected from the structural biology point of view the two distinct intracellular pathways activated upon TLR4 dimerization: the MyD88-dependent pathway and the TRIF/IRF3 pathway leading to interferon production. He discussed the molecular events including phosphorylation and ubiquitination that allow the regulation of the pathways.

Septic shock or excessive inflammation are possibly the most severe outcomes due to inadequate negative regulation of TLR4 signaling leading to excessive pro-inflammatory cytokine production. Similarly, TLR4 excessive stimulation by endogenous molecules derived from necrotic or damaged tissues (danger-associated molecular patterns, DAMPs) has been associated to a wide array of inflammatory and autoimmune diseases, including neuroinflammations and vascular inflammations.

In this perspective, M. Christodoulides reviewed the molecular mechanism of *Neisseria* lipooligosaccharide (LOS) interaction with TLR4 and discussed the role of TLR4 in meningococcal and gonococcal infections in the therapeutic perspective to target *Neisseria* infections with TLR4 antagonists or investigating the role of TLR4 in *Neisseria*-vaccine-induced immune responses.

On the other hand, M. Molteni and C. Rossetti analyzed the main families of endogenous TLR4 stimulators (DAMPs) and discussed DAMP/TLR4 activation mechanisms that very often differ from the well-known direct TLR4/MD-2 binding of bacterial endotoxins.

Another aspect still underestimated relative to TLR4 biology is the crosstalk between TLR4 signaling and metabolic regulations occurring in macrophages and dendritic cells (DCs). TLR4 stimulation of DCs or macrophages results in increased glycolytic activity, an essential process to support their pro-inflammatory functions. L. Perrin-Cocon, A. Aublin-Gex, and V. Lotteau reviewed the molecular mechanisms involved in the modulation of central carbon metabolism, from glycolysis, tricarboxylic acid (TCA) cycle, and oxidative phosphorylation to lipid metabolism, upon TLR4 signaling in macrophages and DCs.

Molecules called microRNAs (miRNA) have been recently described as negative regulators of TLR signaling acting as a break on the pathway while others act as positive regulators and act as an accelerator. There is increasing evidence that TLR4 and miRNA crosstalk ensures the fine-tuning of inflammatory response and subsequent healing occurring in tissues after infection or injury. The expression of miRNAs inside the cell, after TLR4 triggering, critically contributes both to the activation and to the shutdown of immune cell response needed for the termination of the inflammatory process. M. Molteni and C. Rossetti gave insight into the intracellular role of TLR4-miRNA axis in the regulation of inflammatory and anti-inflammatory processes.

Interestingly, TLR4 does not only play a role in inducing inflammation but also in inducing tissue regeneration and healing after inflammatory insult, and it maintains homeostasis in the gut and a state of constant "controlled inflammation" due to the stimulation by commensal bacteria. When there is an imbalance of gut microflora, inflammatory bowel diseases can develop.

This dual role of TLR4 signaling is particularly critical in CNS, as discussed by L. De Filippis and F. Peri. TLR4 is expressed in microglia, which is a master player in neuroinflammatory processes, as well as in neural cells: astrocytes, neurons, oligodendrocytes, neural progenitors (NPC), and neural stem cells (NSC). We have observed that TLR4 stimulation by LPS during differentiation enhances neuronogenic potential of human NSC and favors both neuronal and oligodendroglial survival. Consistent with our data, it has been recently reported that endogenous NSC are actively stimulated to proliferate by TLR4 activation after stroke, thus confirming the relevant role of TLR4 in promoting neurogenesis under non-physiological conditions. Altogether, these results indicate that in a therapeutic perspective, TLR4 activity should not to be turned on or off, but should be finely tuned in order to promote neuroregeneration rather than neurodegeneration and to mediate the development of healing immunomodulation rather than of detrimental neuroinflammation.

Recent insight on the role of TLR4 as mediator of inflammatory response in Alzheimer's (AD) and Parkinson's disease (PD) has been reviewed by C. Balducci and G. Forloni. They reported literature data relating activation of TLR4 and the presence of β amyloid and their own data showing that the memory damage and inflammatory effects obtained by intraventricular application of β amyloid oligomers was antagonized by TLR4 inhibitor and completely abolished in TLR4-knockout mice. They also discuss the recently discovered role of TLR4 in the relationship between gut microbiota dysbiosis and increased risk of developing PD.

M. De Paola presented new data on the potential of Human Induced Pluripotent Stem Cells (IPSC) and Cerebral Organoids as models to study how TLR4 regulates immune cell interactions to orchestrate brain development and reaction to injury. He showed how the development of platforms in which microglia, neurons, and macroglia derived from healthy or diseased subjects grow and mature in a single system allows to demonstrate the role of TLR4 in mediating neuroinflammation.

We thank all authors, that are also our good friends, for having accepted the challenge to compose this multidisciplinary mosaic around TLR4 functions. Their different expertise in the fields of structural biology, biophysics, medicinal chemistry, computational biology, biochemistry, immunology, pharmacology, and medicine made possible the creation of a unique overview on several aspects of physiological and pathological roles of one of the most fascinating and smart molecules to which we have dedicated a large part of our scientific adventure.

Varese, Italy Carlo Rossetti
Milan, Italy Francesco Peri

Contents

About the Editors

Carlo Rossetti was born in Brescia in 1958 and graduated in Biological Sciences in 1982. Currently, he is Associate Professor in Applied Biology at the University of Insubria (Varese, Italy). His research focuses on the role of Toll-like receptors in tissue regeneration and activation of the innate immunity. In 2008, he founded a start-up for the development of TLR4-antagonist natural molecules.

Francesco Peri Full Professor (Organic and Medicinal Chemistry) at the Department of Biotechnology and Biosciences of the University of Milano-Bicocca (Milano, Italy). He holds a permanent professorship at the École Normale Supérieure (ENS) of Lyon (France), where he teaches a course of Medicinal Chemistry (Master level). In 2012, he had a permanent professorship in Organic Chemistry at the University Paris 5; and in 2012, he was for 1 year Visiting Professor at the Department of Chemistry, University of California, Davis (USA), giving graduate courses.

His research activity focuses on medicinal chemistry, bioorganic chemistry, and drug development. New TLR4 modulators have been developed by his group, and novel cardiac drug leads modulating the activity of the Serca2a protein.

He has numerous international and national collaborations, holds national and international grants, and is the president of the MicrobiotaMi association, whose mission is to diffuse the knowledge and the culture related to studies on human microbiota. He is committed to technology transfer and is Delegate of the Rector for the University of Innovation Foundation (www.u4i.it), an organization devoted to the valorization of innovation and to decisional support to University start-ups.

He is founder of the academic spin-off CP2 Biotech, whose mission is the valorization of new drug hits and the development of innovative therapeutic approaches to vaccine adjuvants and inflammatory diseases.

The original version of this book was revised: This book was initially published with incorrect ISSN numbers which has been corrected now. The correction to this book is available at https://doi.org/10.1007/978-3-030-56319-6_12

Chemically Synthesized TLR4 Ligands, Their Immunological Functions, and Potential as Vaccine Adjuvant

Atsushi Shimoyama and Koichi Fukase

Abstract Lipopolysaccharide (LPS), the major glycoconjugates in the outer membrane of Gram-negative bacteria, and its active center glycolipid, lipid A, are recognized by an innate immune system receptor, Toll-like receptor (TLR) 4, and trigger immunostimulatory effects and thus have the potential to act as vaccine adjuvants. Although canonical *Escherichia coli* LPS induces strong inflammation and acts as an endotoxin due to the ability that hyperstimulates the immune system, recent studies revealed that inflammatory activity of lipid A can be attenuated by the structural modification. Here, we discuss the structure–activity relationship of lipid A and the potential as a vaccine adjuvant, and introduce currently used vaccines that contain LPS and lipid A. We also introduce the strategy of safe lipid A adjuvant development based on human symbiotic bacterial lipid As. Finally, we present studies on lipid A–based self-adjuvant strategy and how structural modifications and conjugations can help regulate the adjuvanticity of lipid A.

Keywords Lipid A · Structure–activity relationship · Vaccine adjuvant · Symbiotic bacteria · Self-adjuvant strategy

1 Introduction

Bacterial components have long been known to have immunostimulatory effects [1]. In the past 300 years, bacterial infections have been occasionally reported to reduce tumor size [2]. In the first attempt to demonstrate the use of immunotherapy for cancer, Coley et al. performed antitumor therapy using *Streptococcus pyogenes* and *Serratia marcescens* in 1893. Immunostimulatory effects of dead *Salmonella*

A. Shimoyama · K. Fukase (✉)
Osaka University, Toyonaka, Osaka, Japan
e-mail: koichi@chem.sci.osaka-u.ac.jp

C. Rossetti, F. Peri (eds.), *The Role of Toll-Like Receptor 4 in Infectious and Non Infectious Inflammation*, Progress in Inflammation Research 87,
https://doi.org/10.1007/978-3-030-56319-6_1

typhimurium cells and of *Mycobacterium tuberculosis* have also been reported in 1916 and 1924, respectively. These immunostimulatory effects have been revealed to be triggered by the recognition of molecular patterns characteristic of pathogens and microorganisms by various innate immune receptors in multicellular organisms, which is now widely known as the innate immune system. Since innate immune stimulators activate acquired immune responses, such as antigen-antibody interactions and cell-mediated immunity, several studies are aimed at using innate immune stimulators as adjuvants [3], that is, vaccine ingredients enhancing antibody production.

Lipopolysaccharide (LPS), the major glycoconjugates in the outer membrane of Gram-negative bacteria, and its terminal glycolipid, lipid A (Fig. 1), which is the active principle of LPS, are well-known stimulators of innate immunity. LPS and lipid A induce various immune responses such as cytokine production, nitric oxide production, active oxygen production, leukocyte migration, antibacterial peptide production, and lymphocyte activation, which trigger the host defense system against bacteria. On the other hand, LPS and lipid A also have extremely strong inflammatory effects and are known as endotoxins. Endotoxin is the major contributor to sepsis, and it triggers severe systemic illness that can cause multiple organ failure, low blood pressure, and septic shock [4]. Since the canonical *Escherichia coli* LPS is highly toxic, it has to be modified to attenuate its inflammatory effects and eliminate significant toxicity for application as an adjuvant. Monophosphoryl lipid A (MPL), a lipid A derivative, has already been developed and has been approved as an adjuvant [5]. Here, we introduce the structure–activity relationship of Toll-like receptor (TLR) 4 ligands, especially of lipid A, and the strategy for

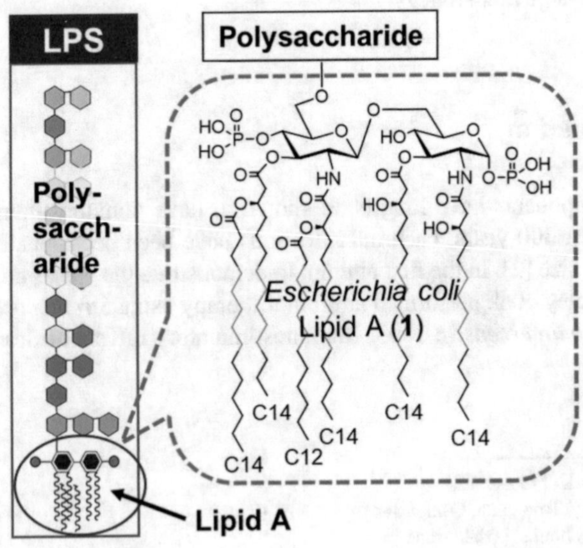

Fig. 1 Bacterial LPS and lipid A

regulating the immune functions of lipid A for the development of lipid A as an adjuvant.

2 Lipid A, a Stimulant of Innate Immunity

In 1892, Pfeiffer showed that *Vibrio cholerae* bacteria produce two different types of toxic components; one is a heat-labile exotoxin and the other is a heat-stable endotoxin [6]. In 1945, Westphal reported that the active component of the endotoxin is LPS, which constitutes the outer membrane of Gram-negative bacteria, and in 1957, lipid A, a glycolipid at the LPS terminus, was reported as the active center of the endotoxin. Shiba and Kusumoto started collaborative research with a German group, and submitted the correct structure of *E. coli* lipid A (**1**) (Fig. 1) and succeeded in the first total synthesis of *E. coli* lipid A (**1**) in 1985, thus confirming that lipid A is the active center of the endotoxin [7–9]. Takayama also identified the lipid A structure at around the same time [10]. Shiba and Kusumoto also accomplished the synthesis of the biosynthetic precursor of *E. coli* lipid A, lipid IVa (**2**) (Fig. 2), and their findings revealed that lipid IVa (**2**) has immunostimulatory effects in mice but has antagonistic effects in humans [11, 12]. At the same time, Golenbock revealed that *Rhodobacter sphaeroides* lipid A (RSLA) (**3**) exhibited antagonistic effects in both humans and mice [13]. These findings suggested the presence of an LPS receptor; therefore, exploratory studies in search of the LPS receptor were conducted. The breakthrough research was accomplished by Hoffmann in 1996. Hoffmann found that the Toll gene of *Drosophila* is essential for the defense mechanism against fungi, which led to the discovery of various innate immune receptors [14]. In 1997, Toll-like receptors (TLRs) were described as human homologues of

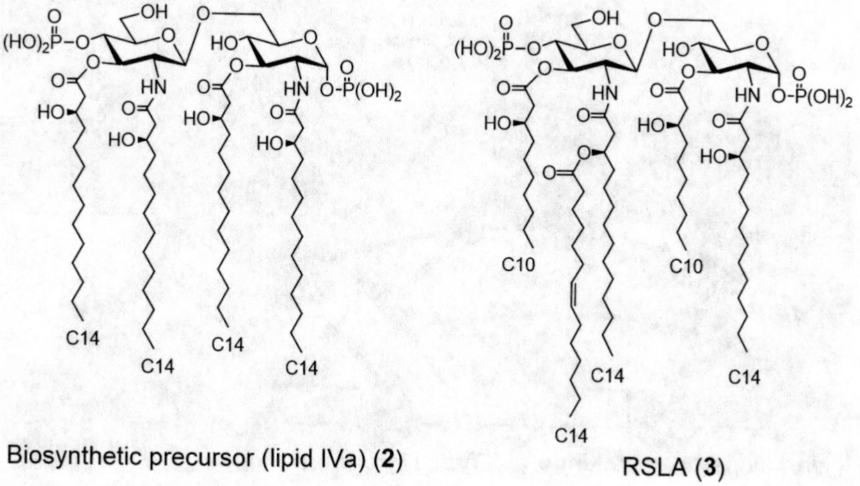

Biosynthetic precursor (lipid IVa) (**2**)

RSLA (**3**)

Fig. 2 Chemical structures of lipid As

the *Drosophila* Toll protein [15], and in 1998 Beutler identified TLR4 as an LPS receptor [16].

TLRs are membrane glycoproteins that contain leucine-rich repeat motifs in the ectodomain and a cytoplasmic signaling domain that is homologous to the interleukin 1 receptor (IL-1R) called TIR (Toll/IL-1R) domain. The TLR4 signal is transmitted via various adaptor molecules, such as MyD88, TIRAP, TRAM, and TRIF, that contain the TIR domain (Fig. 3) [17]. MyD88-mediated signaling activates NF-κB, a transcription factor involved in inflammation, and induces the production of proinflammatory cytokines such as tumor necrosis factor (TNF)-α and IL-6. (Note that cytokine is a general term of proteins secreted by cells and have specific roles in cell–cell interaction and communication.) In particular, inflammatory cytokines are produced by immune cells as a protective response to infection. On the other hand, TRIF-mediated signaling leads to the activation of the interferon (IFN) regulatory factor 3 (IRF3), a transcription factor, and induces the production of the antiviral cytokine type I IFN. Canonical *E. coli* LPS strongly activates both signals (Fig. 3) simultaneously, resulting in a massive inflammatory response, which leads to lethal toxicity. Therefore, regulation of the TLR4 signaling pathway is essential to the development of a lipid A-based adjuvant.

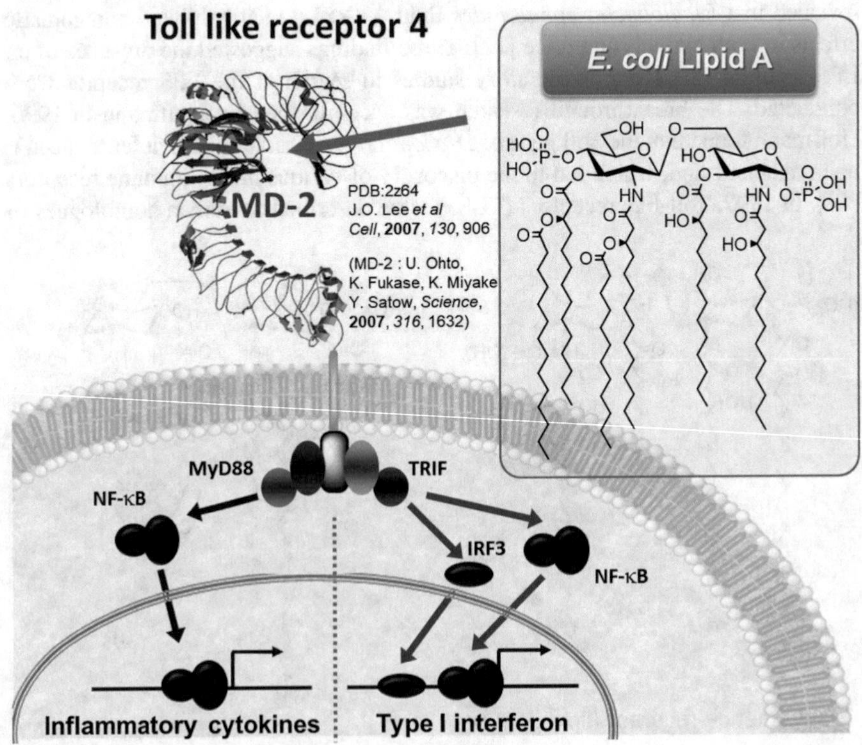

Fig. 3 Innate immune system activation via TLR4/MD2

In order to successfully regulate the TLR4 signaling pathway, understanding the molecular basis of lipid A recognition by TLR4 is necessary. Miyake et al. showed that an accessory protein of TLR4, myeloid differentiation (MD)-2 is essential for TLR4 signal transduction [18]. We accomplished the chemical synthesis of a radio-labeled lipid A analog, and Miyake et al. used it to elucidate the interaction between the TLR4/MD-2 complex and lipid A (Fig. 4) [19]. Miyake et al. also found that the species specificity of TLR4/MD-2 is caused by differences in the recognition of lipid A in MD-2 [20]. Furthermore, X-ray crystal structure studies have led to the understanding of the binding mode between TLR4/MD-2 and its agonists and antagonists. Ohto and Sato et al. revealed the crystal structures of human MD-2 and its complex with lipid IVa [21], and Lee et al. revealed the crystal structure of the mouse TLR4/MD-2 complex with eritoran, a TLR4 antagonist developed by Eisai [22]. In 2009, Lee et al. accomplished the X-ray crystal structure analysis of the

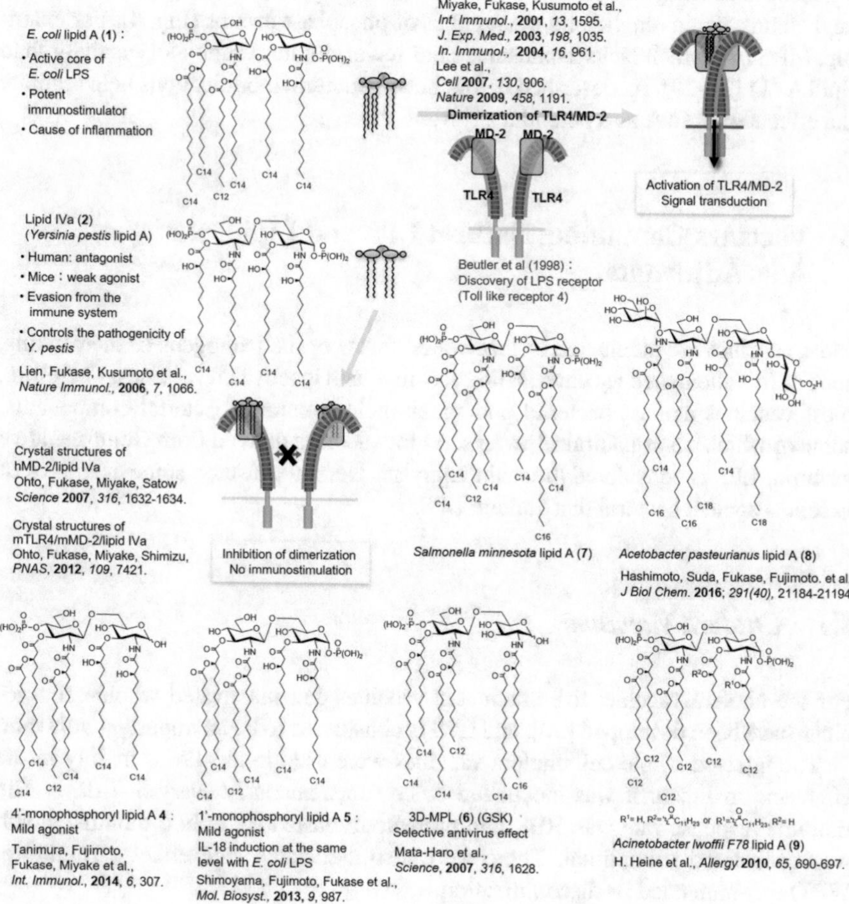

Fig. 4 Chemical structures of lipid As

human TLR4/MD-2 complex with *E. coli* LPS. Five of the six acyl chains of *E. coli* lipid A are housed inside the hydrophobic pocket of MD-2, and the remaining acyl chain interacts with the hydrophobic surface of the adjacent TLR4 [23]. These interactions trigger the dimerization of the TLR4/MD-2 complex to activate the immune response. In the case of an antagonist, the lipid A moiety binds to MD-2 in a form in which lipid A is rotated by 180°, and all acyl chains placed inside MD-2 pocket, which does not cause the dimerization of TLR4/MD-2. Additionally, Ohto et al. revealed the crystal structures of mouse TLR4/MD-2 complex with lipid IVa that acts as an antagonist in humans but as an agonist in mice [24]. In the case of mice, three of the four acyl chains of lipid IVa are housed inside the MD-2 pocket, and the other interacts with the hydrophobic surface of the adjacent TLR4, which leads to the dimerization of TLR4/MD-2. These studies have revealed that differences in the binding mode of lipid A to MD-2 greatly affect TLR4-mediated immune regulation.

Previous structure–activity relationship studies [4, 25, 26] have revealed that agonistic and antagonistic effects can be essentially controlled by the number of acyl chains, chain length, and the number of phosphate groups (Fig. 4). For example, MPL (**4**), which lacks 1-phosphate has lower immunostimulatory activity than lipid A (**1**) [27–29]. As described below, such structural modifications help regulate the effects of lipid A as a potential adjuvant.

3 Vaccines Containing Natural LPS and Lipid A as Adjuvants

Most vaccines are attenuated or inactivated forms of the pathogens or their toxins, such as live attenuated vaccines, killed vaccines, and inactivated vaccines. Therefore, most vaccines against bacterial pathogens include natural bacterial components, some of which act as natural adjuvants. As for vaccines derived from Gram-negative bacteria, LPS is considered the main adjuvant. Here, we discuss some examples of vaccines against bacteria that contain LPS.

3.1 Cholera Vaccines

For the cholera vaccine, live attenuated vaccines and inactivated whole-cell vaccines have been developed [30], and LPS is considered to be an important adjuvant.

The injected whole-cell cholera vaccines were used in the 1960s mainly in the USA and in Japan. It was inoculated twice subcutaneously every 5–7 days. The immune response rate was 50%, and protectivity lasted for only 6 months, which was considered suboptimal. There were also associated side effects. Therefore, WHO recommended its discontinuation.

rBS-WC (commercial name; Dukoral®) is an oral inactivated vaccine licensed in Sweden in 1991 containing killed cholera whole-cell and recombinant cholera toxin B subunit. Now, it is licensed mainly in Europe, Canada, South Asia, and Latin America. The B subunit is the non-toxic subunit of the two proteins that make up the cholera toxin (Ctx). This vaccine has also demonstrated protectivity against pathogenic *E. coli* O139 four months after inoculation. This vaccine has few side effects with an 85–97% effectivity rate. The duration of protection is around 2–3 years.

A live attenuated vaccine has also been developed for cholera, namely, CVD 103-HgR (commercial name; Orochol® or Mutacol®). The vaccine was derived from Inaba strain whose CtxA subunit was deleted. This vaccine was released in the 1990s. The approving countries, effectivity rate, and effectivity period are the same as those of rBS-WC, and only one vaccination is required. However, the manufacturing and sales of these vaccines have been discontinued.

3.2 *Salmonella* Vaccines

Similarly, vaccines for *Salmonella enterica* serovar Typhi include live attenuated, inactivated whole-cell, and subunit vaccines [31].

Vivotif Berna® is an oral live attenuated vaccine developed at the Swiss Serum and Vaccine Institute. It consists of the attenuated typhoid Ty21a strain. This vaccine confers over 2 years of protection and has few side effects. However, it is not recommended for children under 5 years old. It is licensed in Asia, Africa, Europe, America, South America, etc.

Whole-cell inactivated vaccines (heat-phenol inactivated or acetone-inactivated) are mainly used in the USA for patients who cannot receive the orally administered vaccine. It requires two subcutaneous inoculations every 4 weeks, followed by a booster every 3 years. This vaccine provides over 2 years of protection, but it has been reported to have side effects such as fever, headache, general malaise, local swelling, pain, and induration at the inoculation site.

Pasteur Merieux (France) was able to develop a subunit vaccine (commercial name; Typhim Vi®) using the virulence (Vi) capsular polysaccharide antigen purified from *Salmonella* Typhi. Since the Vi antigen has been checked by the endotoxin test, LPS would not be included (on the other hand, a recent study has suggested that the capsular polysaccharide A from the gut commensal *Bacteroides fragilis* included the lipid A structure [32]). This vaccine is effective for 2–3 years with a single intramuscular injection and requires a booster every 2 years. Its side effects are similar to those of the whole cell inactivated vaccines but are relatively mild. It can be stored at room temperature (22 °C) for about 3 years, and it is licensed and used in more than 63 countries in Europe, Africa, Asia, Australia, and the USA.

3.3 Other Vaccines

Inactivated vaccines and subunit vaccines have been developed for *Bordetella pertussis*. They are administered as a diphtheria, tetanus, and pertussis (DTP) combination vaccine or a diphtheria, pertussis, tetanus, and inactivated poliovirus (DPT-IPV) combination vaccine. Most combination vaccines generally use the safer acellular pertussis (aP) vaccine; however, the effective whole-cell vaccine is still also used.

Bexsero®, a serogroup B *meningococcal (MenB)* vaccine, contains the meningococcal outer membrane vesicle (OMV). It is suggested that LPS may act as an adjuvant in OMV-based vaccines [33]. On the other hand, the MenB vaccine Trumenba uses a recombinant lipoprotein, a TLR2 ligand, as antigen [34].

Vaccines for other Gram-negative bacteria, such as those for *Streptococcus pneumoniae*, *Haemophilus influenzae* type b, and *Neisseria meningitidis* (serogroups A, B, C, Y, and W-135), have also been developed, but all of them are purified component vaccines, and so LPS would not be included.

As such, vaccines, including those administered orally, that potentially contain LPS are widely used. As described below, it is reported that some LPS retain their immunostimulatory effects even when administered orally, in which case there are less side effects than injected whole-cell vaccines with LPS.

4 LPS and Lipid A Derived from the Environment and from Fermented Foods as Adjuvant Candidates

Studies have suggested that LPS and lipid As derived from the environment and from fermented foods are involved in immunomodulation, and thus are also potential adjuvant candidates.

Pantoea agglomerans is a Gram-negative bacteria widely found in soil and plants, such as wheat, rice, sweet potato, apples, and pears. It has been detected during the fermentation process of rye bread [35]. Oral administration of *P. agglomerans* LPS has been associated with immunopotentiating effects [36, 37]. In addition, *P. agglomerans* lipid A was found to be a mixture of both the *E. coli* type lipid A (**1**) and *Salmonella minnesota* type lipid A (**7**) [38] (Fig. 4).

Acetobacter spp. is a genus of Gram-negative bacteria used for acetic acid fermentation. Kurozu (black vinegar), an Asian fermented food, contains LPS derived from *Acetobacter* spp. Recent studies have revealed the chemical structure of *Acetobacter pasteurianus* LPS [39] and its lipid A **8** [40] (Fig. 4). *A. pasteurianus* LPS had weaker immunostimulating effects than *E. coli* LPS, but *A. pasteurianus* LPS and lipid A have potential as novel adjuvants due to their safety.

Hygiene hypothesis states that early childhood exposure to environmental microorganisms reduces the risk of developing allergic diseases. *Acinetobacter lwoffii* F78 [41] was found in fodder following an attempt to search for bacteria with allergy-suppressing abilities. *A. lwoffii* LPS selectively induces T helper 1 (Th1) cell-derived cytokines, such as IL-12 and IFN-γ, which have antiallergic effects. Therefore, *A. lwoffii* F78 LPS, which enables selective immune system activation, and its lipid A **9** (Fig. 4) also have potential as novel adjuvants.

5 Synthetic and Semi-synthetic Lipid As as Adjuvants

As mentioned, we have found that MPL **4** (Fig. 4) that lacks 1-phosphate has displayed significantly weaker inflammatory effects than *E. coli* lipid A (**1**) and has exhibited mild immunomodulatory effects [27–29]. This MPL **4** is less dependent on CD14, a glycosylphosphatidylinositol-anchored receptor known to serve as a co-receptor for several TLR4. Additionally, TLR4/MD2 dimerization in response to MPL is greatly reduced compared to the response to *E. coli* LPS. MPL **4** has shown CD14-independent but MyD88-dependent TNFα-producing ability, and TRIF-dependent CD86 upregulating and IFNβ-inducing ability [28].

Similar to **4**, MPL **5**, which lacks 4′-phosphate, also exhibits mild immunomodulative effects. However, while the ability of MPL **4** to induce IL-18 production is lower than that of *E. coli* LPS, and MPL **5** exhibits the same level of IL-18 induction as *E. coli* LPS [29]. Therefore, these MPLs are expected to be developed as future adjuvants with different adjuvant effects.

GlaxoSmithKline developed 3D-MPL (**6**) [5] (Fig. 4) with a 4′-monophosphoryl structure similar to that of MPL **4**. 3D-MPL (**6**) has been successfully attenuated by optimizing the lipid A structure, especially the acyl group and phosphate group. Currently, it is derivatized and produced from *Salmonella minnesota* R595 LPS. This molecule was reported to selectively activate the TRIF-dependent pathway of the two signaling cascades triggered by the TLR4/MD2 complex (Fig. 3) [5]. GlaxoSmithKline has developed the adjuvant AS04, a mixture of 3D-MPL (**6**) and aluminum salts. AS04 induces cell-mediated immune responses and exhibits antiviral effects, and it is used as adjuvant for the human papillomavirus (HPV) vaccine Cervarix and HBV vaccine Fendrix.

In the course of infection with *Plasmodium falciparum*, the malaria-causing agent, infectious sporozoites are infused into the human blood from the salivary glands during blood feeding of the *Anopheles* vector mosquitoes. Therefore, the development of vaccines targeting sporozoite surface proteins was facilitated. GlaxoSmithKline has developed a malaria vaccine candidate RTS,S/AS01, which is currently in Phase III clinical trials. Recombinant protein RTS,S consists of a segment of a sporozoite protein and the hepatitis B virus surface antigen, and AS01 is a liposome adjuvant composed of cholesterol, MPL, and QS21 (a saponin derived from a South American native tree *Kiraja saponaria*). GlaxoSmithKline has also developed the adjuvant AS02 (composed of oil emulsion, squalene, QS21, and MPL). A malaria vaccine RTS,S/AS02 and the herpes zoster vaccine HZ/su that use AS01 as adjuvant are currently in Phase III clinical trials.

Fig. 5 RC-529

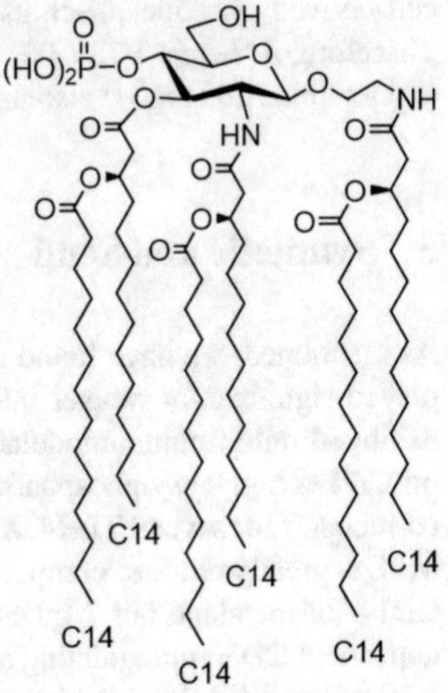

Additionally, an MPL mimic, RC-529 (**10**) (Fig. 5), was approved in Argentina in 2003 as adjuvant for the hepatitis B virus vaccine.

Lipid A derivative adjuvants, such as MPL, can induce anti-inflammatory cytokines such as IL-10 while regulating the induction of inflammatory cytokines such as IL-6 [42]. Therefore, compared with other adjuvants that do not induce IL-10, lipid A derivatives are believed to pose lower risk for the development of adjuvant-induced autoimmune diseases. As such, further development of lipid A adjuvants is expected.

6 Developing Lipid A Adjuvant Candidates

Several studies have reported that some symbiotic and pathogenic bacteria evade the innate immune system like stealth aircrafts. Plague is a rodent infection caused by *Yersinia pestis*, but it also infects humans mainly through fleas. *Y. pestis* produces a protein toxin that destroys peripheral blood vessels and triggers edema and necrosis. *Y. pestis* possesses lipid A that has the same structure as *E. coli* lipid A (**1**) at 27 °C, but its structure changes to the antagonistic lipid IVa (**2**) in the mammalian body temperature of 37 °C (Fig. 4) [43]. The antagonistic effect of *Y. pestis* LPS would decrease the host's innate immune response, which contributes to the virulence of *Y. pestis*.

To investigate the chemical communications between bacteria and host, we recently focused on lipid As from human symbiotic bacteria, and revealed a close relationship between bacterial characteristics and its lipid A activity [29, 44, 45]. Extracted LPS from parasitic bacteria such as *Helicobacter pylori*, which inhabits the stomach and causes gastric ulcer, and *Porphynomonas gingivalis*, an oral bacteria and one of the causes of periodontal disease, exhibited weak immunostimulatory effects. Their LPS have been reported to be associated with the development of chronic inflammation and atherosclerosis [46–49]. *H. pylori* and *P. gingivalis* lipid As are known to have characteristic and heterogeneous structures, and it was suggested that the ability of their LPS to regulate TLR4/MD2 is a factor for the specific biological activity of these parasitic bacterial LPS [50]. *E. coli* lipid A (**1**) (Fig. 1) contains six fatty chains (C12-C14). On the other hand, *H. pylori* lipid A **11, 12** contains a smaller number of fatty chains (three to four), but the chain length is longer (C16-C18) than that of the *E. coli* lipid A (Fig. 6). Regarding the phosphate group, *E. coli* lipid A (**1**) contains two phosphate groups at the 1- and 4′-position, while *H. pylori* lipid As **11a, 12a** contain a phosphate group only at the 1-position, and *H. pylori* lipid As **11b, 12b** have an ethanolamine condensed phosphate group only at the 1-position. Meanwhile, *P. gingivalis* lipid As **14–17** possess three to five fatty chains (C15 ~ C17), including chains with terminal branching. Similar to the *H. pylori* lipid A, *P. gingivalis* lipid As **14–17** also contain phosphate groups only at the 1-position. Thus, the structural features common to the parasitic bacterial lipid As **11, 12, 14–16** are: fatty chains are longer and more diverse than those in *E. coli* lipid A (**1**), and only 1-position is phosphorylated. It also means that the parasitic bacterial lipid A has an MPL structure that is similar to the 3D-MPL (**6**), which has only one phosphate group at the 4′-position.

We have accomplished comprehensive chemical synthesis of parasitic bacterial partial structures **11–17** and evaluated their immunological functions. The

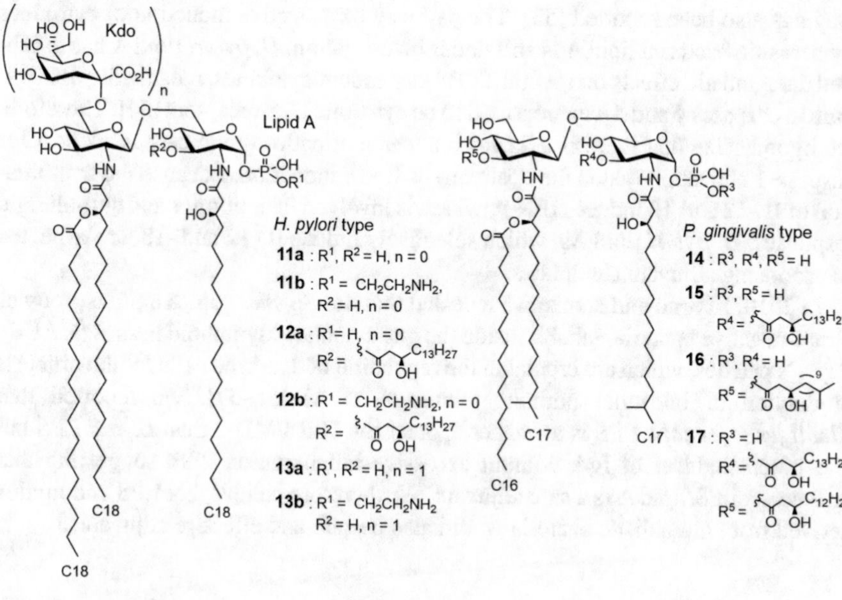

Kdo
Lipid A
H. pylori type
11a : R^1, R^2 = H, n = 0
11b : R^1 = $CH_2CH_2NH_2$, R^2 = H, n = 0
12a : R^1 = H, n = 0, R^2 = (3-hydroxy acyl, $C_{13}H_{27}$)
12b : R^1 = $CH_2CH_2NH_2$, n = 0, R^2 = (3-hydroxy acyl, $C_{13}H_{27}$)
13a : R^1, R^2 = H, n = 1
13b : R^1 = $CH_2CH_2NH_2$, R^2 = H, n = 1
P. gingivalis type
14 : R^3, R^4, R^5 = H
15 : R^3, R^5 = H, R^4 = (3-hydroxy acyl, $C_{13}H_{27}$)
16 : R^3, R^4 = H, R^5 = (branched 3-hydroxy acyl)
17 : R^3 = H, R^4 = (3-hydroxy acyl, $C_{13}H_{27}$), R^5 = (3-hydroxy acyl, $C_{12}H_{25}$)

Fig. 6 Chemical structures of parasitic bacterial LPS partial structures

cytokine-inducing activities of chemically synthesized **11–17** in human peripheral whole blood were measured and compared with that of *E. coli* LPS. Their antagonistic effects were evaluated by competition assays with *E. coli* LPS. Parasitic bacterial lipid A **11a**, **12a**, **14–15** having three to four fatty acid chains and one normal phosphate group showed antagonistic effects on the induction of proinflammatory cytokines such as IL-6 and TNF-α. On the other hand, *H. pylori* lipid A **11b**, **12b**, having three to four fatty acid chains and one ethanolamine phosphate group, and *P. gingivalis* lipid A **16**, having five fatty acid chains and one normal phosphate group, were able to induce the production of proinflammatory cytokines such as IL-6 and TNF-α but to a great lesser degree compared to *E. coli* LPS. In the natural LPS, lipid A is linked to the terminus of the polysaccharide part through a unique acidic sugar 2-keto-3-deoxy-D-manno-ocutulosonic acid (Kdo). The introduction of Kdo to *E. coli* lipid A enhanced its immunostimulatory effects [51]. On the other hand, for *H. pylori* lipid A, **13a**, which has Kdo added to **11a**, showed stronger antagonistic effects than the unmodified **11a**; **13b**, which has a Kdo added to **11b**, switched to antagonist. Therefore, our studies have revealed that in *H. pylori* LPS, the active principal is Kdo-lipid A not lipid A itself. These results suggest that parasitic bacteria are evolving to escape the host's innate immune responses, and their LPS/lipid A show antagonistic or extremely weak agonistic effects to favor infection to the host. As such, we have found a close relationship between bacterial characteristics and their lipid A activity.

All parasitic bacterial lipid A **11–17** were found to induce IL-12 and -18, which are involved in chronic inflammation. In particular, we have found that **11a**, **12a**, **13–15** selectively induce IL-12 and -18. IL-18 induction triggered by LPS was reported to be dependent on the TRIF pathway [52], but a TRIF-independent pathway has also been reported [53]. The pathway for selective induction of cytokines by parasitic bacterial lipid A is still under investigation. *H. pylori* lipid A has exhibited antagonistic effects on several TLR4-dependent cytokines, meanwhile in 2014, human caspases 4 and 5 were reported to be cytosolic LPS receptors [54]. Therefore, we hypothesize that there is a TLR4-independent pathway via caspase 4 or 5 for caspase-1 activation that is the upstream of IL-18 induction. Because the combination of IL-12 and 18 induces IFN-γ, which is involved in antitumor and anti-allergic responses, *H. pylori* lipid As, which selectively induce IL-12 and -18, are expected as promising adjuvant candidates.

In 2010, Kiyono and Kunizawa revealed that *Alcaligenes* spp., a unique group of Gram-negative bacteria, inhabit inside the gut-associated lymphoid tissues (GALT), Peyer's patches, which are crucial in the regulation of dendritic cells for the efficient production of intestinal immunoglobulin A (IgA) [55–57]. We reported that *Alcaligenes faecalis* LPS is a weaker agonist for TLR4/MD-2 than *E. coli* LPS but is a potent inducer of IgA without excessive inflammation [58], suggesting that *A. faecalis* LPS/lipid A is a safe adjuvant. We therefore believe that LPS and lipid A derived from mutualistic bacteria would also be safe and effective adjuvants.

7 Lipid A Based Self-adjuvant Vaccines

The self-adjuvant strategy, which promotes a more efficient antibody production by covalently binding an antigen and an adjuvant, has recently become a focus of several studies, especially the use of a lipopeptide adjuvant (TLR2 ligand) [6, 59–67]. The antigen-adjuvant complex is actively taken up by dendritic cells via the innate immune ligand (adjuvant), and the adjuvant activates the immune system to induce cytokine production and efficient antibody production. The advantageous feature of this strategy is that an antigen and an adjuvant covalently linked are taken up by the same dendritic cell, which can induce an immune response to the antigen specifically (Fig. 7).

Lipid A-based self-adjuvant vaccines have also been developed. Guo et al. synthesized the adjuvant-antigen complex (Fig. 8a) in which the *E. coli* type MPL and α-2,9-oligosialic acid (meningococcal antigen) were bound via a linker, and enhanced antibody production (especially of IgG2b and 2c) has been observed [68]. Jiang et al. synthesized the adjuvant-antigen complex (Fig. 8b) in which RC-529 (**10**) and the Thomsen–Friedenreich (TF) antigen (a tumor-associated carbohydrate antigen) were bound via a linker [69].

Self-adjuvant vaccines that combine an innate immune ligand (adjuvant) and an antigen, such as the type B meningococcal vaccine Trumenba, not only have high potency but are also excellent in terms of quality retention and safety control because it is easy to obtain high purity products. For lipid A–based self-adjuvant vaccines, structural modifications sometimes abolish the ability of lipid A to trigger innate immune responses, and the development of active lipid A–antigen complex can be challenging. But once a simple and universal conjugation method that can retain the lipid A function has been developed, it can then be used for the development of various synthetic vaccines and will therefore be of great contribution to immunology.

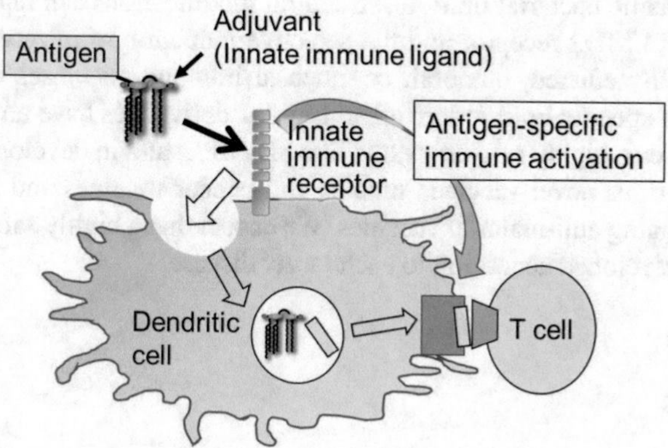

Fig. 7 Self-adjuvant strategy

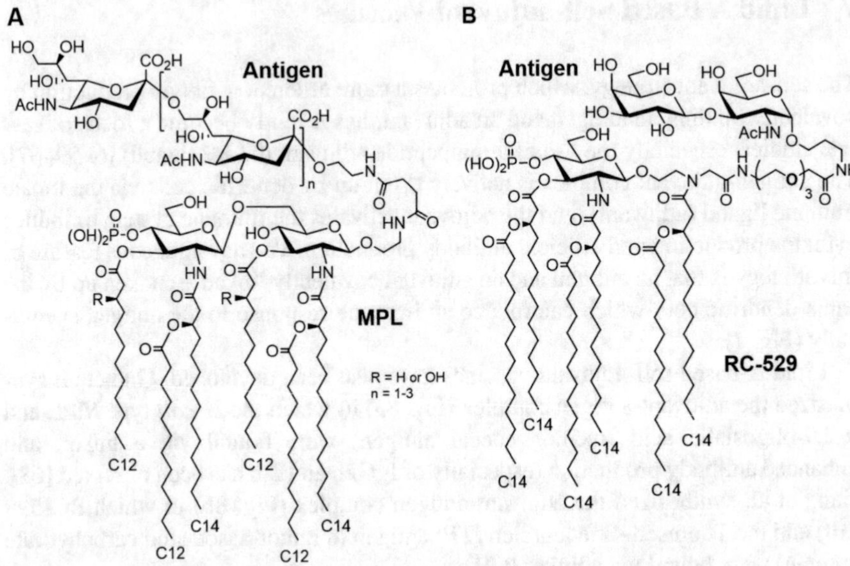

Fig. 8 Chemical structures of adjuvant-antigen complexes: (**a**) *E. coli* type MPL conjugated with α-2,9-oligosialic acid (meningococcal antigen), (**b**) RC-529 (10) conjugated with the Thomsen–Friedenreich (TF) antigen (a tumor-associated carbohydrate antigen)

8 Conclusions

We have discussed the structure–activity relationship of TLR4 ligands, especially of lipid A, and the strategy for regulating the immune functions of lipid A for the development of lipid A as an adjuvant.

A delicate balance between the volumes of the hydrophilic and hydrophobic parts and the spatial arrangement of the hydrophobic and acidic functional groups are important for the expression and control of lipid A activity. As in the case of MPL and parasitic bacterial lipid As, structural modifications can regulate the activation of TLR4/MD2 receptor and the selective induction of intracellular signals. Therefore, cell-mediated, humoral, or mucosal immune responses could be controlled using a specific lipid A derivative. Lipid A derivatives have already been put to practical use as highly safe adjuvants like 3D-MPL, and in development as components of various novel vaccines such as anti-cancer vaccines and anti-protozoal vaccines including anti-malarial vaccines. We expect that a highly safe lipid A adjuvant will be developed according to each target disease.

References

1. Kusumoto S, Fukase K, Shiba T. Key structures of bacterial peptidoglycan and lipopolysaccharide triggering the innate immune system of higher animals: chemical synthesis and functional studies. Proc Jpn Acad Ser B Phys Biol Sci. 2010;86(4):322–37.

2. Wei MQ, Mengesha A, Good D, Anne J. Bacterial targeted tumour therapy-dawn of a new era. Cancer Lett. 2008;259(1):16–27. https://doi.org/10.1016/j.canlet.2007.10.034.
3. Leroux-Roels G. Unmet needs in modern vaccinology: adjuvants to improve the immune response. Vaccine. 2010;28(Suppl 3):C25–36. https://doi.org/10.1016/j.vaccine.2010.07.021.
4. Molinaro A, Holst O, Di Lorenzo F, Callaghan M, Nurisso A, D'Errico G, Zamyatina A, Peri F, Berisio R, Jerala R, Jimenez-Barbero J, Silipo A, Martin-Santamaria S. Chemistry of lipid A: at the heart of innate immunity. Chem Eur J. 2015;21(2):500–19. https://doi.org/10.1002/chem.201403923.
5. Mata-Haro V, Cekic C, Martin M, Chilton PM, Casella CR, Mitchell TC. The vaccine adjuvant monophosphoryl lipid A as a TRIF-biased agonist of TLR4. Science. 2007;316(5831):1628–32. https://doi.org/10.1126/science.1138963.
6. Rietschel ET, Westphal O. Endotoxin: historical perspectives. In: Endotoxin in health and disease. New York: Marcel Dekker; 1999.
7. Imoto M, Kusumoto S, Shiba T, Naoki H, Iwashita T, Rietschel ET, Wollenweber HW, Galanos C, Luderitz O. Chemical-structure of Escherichia-Coli lipid-a – linkage site of acyl-groups in the disaccharide backbone. Tetrahedron Lett. 1983;24(37):4017–20. https://doi.org/10.1016/S0040-4039(00)88251-9.
8. Imoto M, Kusumoto S, Shiba T, Rietschel ET, Galanos C, Luederitz O. Chemical structure of Escherichia coli lipid A. Tetrahedron Lett. 1985;26(7):907–8.
9. Imoto M, Yoshimura H, Shimamoto T, Sakaguchi N, Kusumoto S, Shiba T. Total synthesis of Escherichia coli lipid A, the endotoxically active principle of cell-surface lipopolysaccharide. Bull Chem Soc Jpn. 1987;60(6):2205–14.
10. Takayama K, Qureshi N, Mascagni P. Complete structure of lipid A obtained from the lipopolysaccharides of the heptoseless mutant of Salmonella typhimurium. J Biol Chem. 1983;258(21):12801–3.
11. Flad HD, Loppnow H, Feist W, Wang MH, Brade H, Kusumoto S, Rietschel ET, Ulmer AJ. Interleukin 1 and tumor necrosis factor: studies on the induction by lipopolysaccharide partial structures. Lymphokine Res. 1989;8(3):235–8.
12. Wang MH, Feist W, Herzbeck H, Brade H, Kusumoto S, Rietschel ET, Flad HD, Ulmer AJ. Suppressive effect of lipid A partial structures on lipopolysaccharide or lipid A-induced release of interleukin 1 by human monocytes. FEMS Microbiol Immunol. 1990;2(3):179–85. https://doi.org/10.1111/j.1574-6968.1990.tb03517.x.
13. Golenbock DT, Hampton RY, Qureshi N, Takayama K, Raetz CR. Lipid A-like molecules that antagonize the effects of endotoxins on human monocytes. J Biol Chem. 1991;266(29):19490–8.
14. Lemaitre B, Nicolas E, Michaut L, Reichhart JM, Hoffmann JA. The dorsoventral regulatory gene cassette spatzle/Toll/cactus controls the potent antifungal response in Drosophila adults. Cell. 1996;86(6):973–83. https://doi.org/10.1016/s0092-8674(00)80172-5.
15. Medzhitov R, Preston-Hurlburt P, Janeway CA Jr. A human homologue of the Drosophila Toll protein signals activation of adaptive immunity. Nature. 1997;388(6640):394–7. https://doi.org/10.1038/41131.
16. Poltorak A, He X, Smirnova I, Liu MY, Van Huffel C, Du X, Birdwell D, Alejos E, Silva M, Galanos C, Freudenberg M, Ricciardi-Castagnoli P, Layton B, Beutler B. Defective LPS signaling in C3H/HeJ and C57BL/10ScCr mice: mutations in Tlr4 gene. Science. 1998;282(5396):2085–8. https://doi.org/10.1126/science.282.5396.2085.
17. Kawai T, Akira S. The role of pattern-recognition receptors in innate immunity: update on Toll-like receptors. Nat Immunol. 2010;11(5):373–84. https://doi.org/10.1038/ni.1863.
18. Shimazu R, Akashi S, Ogata H, Nagai Y, Fukudome K, Miyake K, Kimoto M. MD-2, a molecule that confers lipopolysaccharide responsiveness on Toll-like receptor 4. J Exp Med. 1999;189(11):1777–82. https://doi.org/10.1084/jem.189.11.1777.
19. Akashi S, Saitoh S, Wakabayashi Y, Kikuchi T, Takamura N, Nagai Y, Kusumoto Y, Fukase K, Kusumoto S, Adachi Y, Kosugi A, Miyake K. Lipopolysaccharide interaction with cell surface Toll-like receptor 4-MD-2: higher affinity than that with MD-2 or CD14. J Exp Med. 2003;198(7):1035–42. https://doi.org/10.1084/jem.20031076.

20. Akashi S, Nagai Y, Ogata H, Oikawa M, Fukase K, Kusumoto S, Kawasaki K, Nishijima M, Hayashi S, Kimoto M, Miyake K. Human MD-2 confers on mouse Toll-like receptor 4 species-specific lipopolysaccharide recognition. Int Immunol. 2001;13(12):1595–9. https://doi.org/10.1093/intimm/13.12.1595.
21. Ohto U, Fukase K, Miyake K, Satow Y. Crystal structures of human MD-2 and its complex with antiendotoxic lipid IVa. Science. 2007;316(5831):1632–4. https://doi.org/10.1126/science.1139111.
22. Kim HM, Park BS, Kim JI, Kim SE, Lee J, Oh SC, Enkhbayar P, Matsushima N, Lee H, Yoo OJ, Lee JO. Crystal structure of the TLR4-MD-2 complex with bound endotoxin antagonist Eritoran. Cell. 2007;130(5):906–17. https://doi.org/10.1016/j.cell.2007.08.002.
23. Park BS, Song DH, Kim HM, Choi BS, Lee H, Lee JO. The structural basis of lipopolysaccharide recognition by the TLR4-MD-2 complex. Nature. 2009;458(7242):1191–5. https://doi.org/10.1038/nature07830.
24. Ohto U, Fukase K, Miyake K, Shimizu T. Structural basis of species-specific endotoxin sensing by innate immune receptor TLR4/MD-2. Proc Natl Acad Sci U S A. 2012;109(19):7421–6. https://doi.org/10.1073/pnas.1201193109.
25. Fukase KF, Fujimoto Y, Shimoyama A, Tanaka K. Synthesis of bacterial Glycoconjugates and their bio-functional studies in innate immunity. J Synth Org Chem Jpn. 2012;70(2):113–30.
26. Kusumoto S, Fukase K. Synthesis of endotoxic principle of bacterial lipopolysaccharide and its recognition by innate immune system of hosts. Chem Rec. 2006;6(6):333–43. https://doi.org/10.1002/tcr.20098.
27. Brade L, Brandenburg K, Kuhn HM, Kusumoto S, Macher I, Rietschel ET, Brade H. The immunogenicity and antigenicity of lipid A are influenced by its physicochemical state and environment. Infect Immun. 1987;55(11):2636–44.
28. Tanimura N, Saitoh S, Ohto U, Akashi-Takamura S, Fujimoto Y, Fukase K, Shimizu T, Miyake K. The attenuated inflammation of MPL is due to the lack of CD14-dependent tight dimerization of the TLR4/MD2 complex at the plasma membrane. Int Immunol. 2014;26(6):307–14. https://doi.org/10.1093/intimm/dxt071.
29. Fujimoto Y, Shimoyama A, Saeki A, Kitayama N, Kasamatsu C, Tsutsui H, Fukase K. Innate immunomodulation by lipophilic termini of lipopolysaccharide; synthesis of lipid As from Porphyromonas gingivalis and other bacteria and their immunomodulative responses. Mol BioSyst. 2013;9(5):987–96. https://doi.org/10.1039/c3mb25477a.
30. Ryan ET, Calderwood SB. Cholera vaccines. Clin Infect Dis. 2000;31(2):561–5. https://doi.org/10.1086/313951.
31. Keystone J. Typhoid vaccination – update. Can J Infect Dis = J Can des Mal Infect. 1995;6(5):231. https://doi.org/10.1155/1995/919582.
32. Erturk-Hasdemir D, Oh SF, Okan NA, Stefanetti G, Gazzaniga FS, Seeberger PH, Plevy SE, Kasper DL. Symbionts exploit complex signaling to educate the immune system. Proc Natl Acad Sci U S A. 2019;116(52):26157–66. https://doi.org/10.1073/pnas.1915978116.
33. Watson PS, Turner DP. Clinical experience with the meningococcal B vaccine, Bexsero((R)): prospects for reducing the burden of meningococcal serogroup B disease. Vaccine. 2016;34(7):875–80. https://doi.org/10.1016/j.vaccine.2015.11.057.
34. Luo Y, Friese OV, Runnels HA, Khandke L, Zlotnick G, Aulabaugh A, Gore T, Vidunas E, Raso SW, Novikova E, Byrne E, Schlittler M, Stano D, Dufield RL, Kumar S, Anderson AS, Jansen KU, Rouse JC. The dual role of lipids of the lipoproteins in Trumenba, a self-Adjuvanting vaccine against meningococcal meningitis B disease. AAPS J. 2016;18(6):1562–75. https://doi.org/10.1208/s12248-016-9979-x.
35. Kariluoto S, Aittamaa M, Korhola M, Salovaara H, Vahteristo L, Piironen V. Effects of yeasts and bacteria on the levels of folates in rye sourdoughs. Int J Food Microbiol. 2006;106(2):137–43. https://doi.org/10.1016/j.ijfoodmicro.2005.06.013.
36. Dutkiewicz J, Mackiewicz B, Lemieszek MK, Golec M, Milanowski J. Pantoea agglomerans: a mysterious bacterium of evil and good. Part IV. Beneficial effects. Ann Agric Environ Med. 2016;23(2):206–22. https://doi.org/10.5604/12321966.1203879.

37. Hebishima T, Matsumoto Y, Watanabe G, Soma G, Kohchi C, Taya K, Hayashi Y, Hirota Y. Oral administration of immunopotentiator from Pantoea agglomerans 1 (IP-PA1) improves the survival of B16 melanoma-inoculated model mice. Exp Anim. 2011;60(2):101–9.
38. Tsukioka D, Nishizawa T, Miyase T, Achiwa K, Suda T, Soma G, Mizuno D. Structural characterization of lipid A obtained from Pantoea agglomerans lipopolysaccharide. FEMS Microbiol Lett. 1997;149(2):239–44.
39. Pallach M, Di Lorenzo F, Facchini FA, Gully D, Giraud E, Peri F, Duda KA, Molinaro A, Silipo A. Structure and inflammatory activity of the LPS isolated from Acetobacter pasteurianus CIP103108. Int J Biol Macromol. 2018;119:1027–35. https://doi.org/10.1016/j.ijbiomac.2018.08.035.
40. Hashimoto M, Ozono M, Furuyashiki M, Baba R, Hashiguchi S, Suda Y, Fukase K, Fujimoto Y. Characterization of a novel d-Glycero-d-talo-oct-2-ulosonic acid-substituted lipid A moiety in the lipopolysaccharide produced by the acetic acid bacterium Acetobacter pasteurianus NBRC 3283. J Biol Chem. 2016;291(40):21184–94. https://doi.org/10.1074/jbc.M116.751271.
41. Debarry J, Hanuszkiewicz A, Stein K, Holst O, Heine H. The allergy-protective properties of Acinetobacter lwoffii F78 are imparted by its lipopolysaccharide. Allergy. 2010;65(6):690–7. https://doi.org/10.1111/j.1398-9995.2009.02253.x.
42. Martin M, Michalek SM, Katz J. Role of innate immune factors in the adjuvant activity of monophosphoryl lipid A. Infect Immun. 2003;71(5):2498–507.
43. Kawahara K, Tsukano H, Watanabe H, Lindner B, Matsuura M. Modification of the structure and activity of lipid A in Yersinia pestis lipopolysaccharide by growth temperature. Infect Immun. 2002;70(8):4092–8.
44. Shimoyama A, Saeki A, Tanimura N, Tsutsui H, Miyake K, Suda Y, Fujimoto Y, Fukase K. Chemical synthesis of Helicobacter pylori lipopolysaccharide partial structures and their selective proinflammatory responses. Chem Eur J. 2011;17(51):14464–74. https://doi.org/10.1002/chem.201003581.
45. Fujimoto Y, Shimoyama A, Suda Y, Fukase K. Synthesis and immunomodulatory activities of Helicobacter pylori lipophilic terminus of lipopolysaccharide including lipid A. Carbohydr Res. 2012;356:37–43. https://doi.org/10.1016/j.carres.2012.03.013.
46. Hynes SO, Ferris JA, Szponar B, Wadstrom T, Fox JG, O'Rourke J, Larsson L, Yaquian E, Ljungh A, Clyne M, Andersen LP, Moran AP. Comparative chemical and biological characterization of the lipopolysaccharides of gastric and enterohepatic helicobacters. Helicobacter. 2004;9(4):313–23. https://doi.org/10.1111/j.1083-4389.2004.00237.x.
47. Nielsen H, Birkholz S, Andersen LP, Moran AP. Neutrophil activation by helicobacter pylori lipopolysaccharides. J Infect Dis. 1994;170(1):135–9.
48. Perez-Perez GI, Shepherd VL, Morrow JD, Blaser MJ. Activation of human THP-1 cells and rat bone marrow-derived macrophages by Helicobacter pylori lipopolysaccharide. Infect Immun. 1995;63(4):1183–7.
49. Danesh J, Wong Y, Ward M, Muir J. Chronic infection with Helicobacter pylori, Chlamydia pneumoniae, or cytomegalovirus: population based study of coronary heart disease. Heart. 1999;81(3):245–7.
50. Triantafilou M, Gamper FG, Lepper PM, Mouratis MA, Schumann C, Harokopakis E, Schifferle RE, Hajishengallis G, Triantafilou K. Lipopolysaccharides from atherosclerosis-associated bacteria antagonize TLR4, induce formation of TLR2/1/CD36 complexes in lipid rafts and trigger TLR2-induced inflammatory responses in human vascular endothelial cells. Cell Microbiol. 2007;9(8):2030–9. https://doi.org/10.1111/j.1462-5822.2007.00935.x.
51. Yoshizaki H, Fukuda N, Sato K, Oikawa M, Fukase K, Suda Y, Kusumoto S. First Total Synthesis of the Re-Type Lipopolysaccharide This work was supported by the Research for the Future Program (No. 97L00502) from the Japan Society for the Promotion of Science. H.Y. is grateful for a JSPS Research Fellowship for Young Scientists (No. 1241) from the Japan Society for the Promotion of Science. The authors are grateful to Mr. Seiji Adachi for his skillful measurement of NMR spectra. Angew Chem Int Ed Engl. 2001;40(8):1475–80.

52. Imamura M, Tsutsui H, Yasuda K, Uchiyama R, Yumikura-Futatsugi S, Mitani K, Hayashi S, Akira S, Taniguchi S, Van Rooijen N, Tschopp J, Yamamoto T, Fujimoto J, Nakanishi K. Contribution of TIR domain-containing adapter inducing IFN-beta-mediated IL-18 release to LPS-induced liver injury in mice. J Hepatol. 2009;51(2):333–41. https://doi.org/10.1016/j.jhep.2009.03.027.
53. Kanneganti TD, Lamkanfi M, Kim YG, Chen G, Park JH, Franchi L, Vandenabeele P, Nunez G. Pannexin-1-mediated recognition of bacterial molecules activates the cryopyrin inflammasome independent of Toll-like receptor signaling. Immunity. 2007;26(4):433–43. https://doi.org/10.1016/j.immuni.2007.03.008.
54. Shi J, Zhao Y, Wang Y, Gao W, Ding J, Li P, Hu L, Shao F. Inflammatory caspases are innate immune receptors for intracellular LPS. Nature. 2014;514(7521):187–92. https://doi.org/10.1038/nature13683.
55. Obata T, Goto Y, Kunisawa J, Sato S, Sakamoto M, Setoyama H, Matsuki T, Nonaka K, Shibata N, Gohda M, Kagiyama Y, Nochi T, Yuki Y, Fukuyama Y, Mukai A, Shinzaki S, Fujihashi K, Sasakawa C, Iijima H, Goto M, Umesaki Y, Benno Y, Kiyono H. Indigenous opportunistic bacteria inhabit mammalian gut-associated lymphoid tissues and share a mucosal antibody-mediated symbiosis. Proc Natl Acad Sci U S A. 2010;107(16):7419–24. https://doi.org/10.1073/pnas.1001061107.
56. Fung TC, Bessman NJ, Hepworth MR, Kumar N, Shibata N, Kobuley D, Wang K, Ziegler CGK, Goc J, Shima T, Umesaki Y, Sartor RB, Sullivan KV, Lawley TD, Kunisawa J, Kiyono H, Sonnenberg GF. Lymphoid-tissue-resident commensal bacteria promote members of the IL-10 cytokine family to establish mutualism. Immunity. 2016;44(3):634–46. https://doi.org/10.1016/j.immuni.2016.02.019.
57. Sonnenberg GF, Monticelli LA, Alenghat T, Fung TC, Hutnick NA, Kunisawa J, Shibata N, Grunberg S, Sinha R, Zahm AM, Tardif MR, Sathaliyawala T, Kubota M, Farber DL, Collman RG, Shaked A, Fouser LA, Weiner DB, Tessier PA, Friedman JR, Kiyono H, Bushman FD, Chang KM, Artis D. Innate lymphoid cells promote anatomical containment of lymphoid-resident commensal bacteria. Science. 2012;336(6086):1321–5. https://doi.org/10.1126/science.1222551.
58. Shibata N, Kunisawa J, Hosomi K, Fujimoto Y, Mizote K, Kitayama N, Shimoyama A, Mimuro H, Sato S, Kishishita N, Ishii KJ, Fukase K, Kiyono H. Lymphoid tissue-resident Alcaligenes LPS induces IgA production without excessive inflammatory responses via weak TLR4 agonist activity. Mucosal Immunol. 2018;11(3):693–702. https://doi.org/10.1038/mi.2017.103.
59. Ingale S, Wolfert MA, Gaekwad J, Buskas T, Boons G-J. Robust immune responses elicited by a fully synthetic three-component vaccine. Nat Chem Biol. 2007;3(10):663–7. http://www.nature.com/nchembio/journal/v3/n10/suppinfo/nchembio.2007.25_S1.html
60. Khan S, Weterings JJ, Britten CM, de Jong AR, Graafland D, Melief CJM, van der Burg SH, van der Marel G, Overkleeft HS, Filippov DV, Ossendorp F. Chirality of TLR-2 ligand Pam3CysSK4 in fully synthetic peptide conjugates critically influences the induction of specific CD8+ T-cells. Mol Immunol. 2009;46(6):1084–91. https://doi.org/10.1016/j.molimm.2008.10.006.
61. Kaiser A, Gaidzik N, Becker T, Menge C, Groh K, Cai H, Li Y-M, Gerlitzki B, Schmitt E, Kunz H. Fully synthetic vaccines consisting of tumor-associated MUC1 Glycopeptides and a Lipopeptide ligand of the toll-like receptor 2. Angew Chem Int Ed Engl. 2010;49(21):3688–92. https://doi.org/10.1002/anie.201000462.
62. Wilkinson BL, Day S, Malins LR, Apostolopoulos V, Payne RJ. Self-Adjuvanting multicomponent Cancer vaccine candidates combining per-glycosylated MUC1 Glycopeptides and the toll-like receptor 2 agonist Pam3CysSer. Angew Chem Int Ed Engl. 2011;50(7):1635–9. https://doi.org/10.1002/anie.201006115.
63. Wilkinson BL, Day S, Chapman R, Perrier S, Apostolopoulos V, Payne RJ. Synthesis and immunological evaluation of self-assembling and self-Adjuvanting Tricomponent Glycopeptide Cancer-vaccine candidates. Chem Eur J. 2012;18(51):16540–8. https://doi.org/10.1002/chem.201202629.

64. Lakshminarayanan V, Thompson P, Wolfert MA, Buskas T, Bradley JM, Pathangey LB, Madsen CS, Cohen PA, Gendler SJ, Boons GJ. Immune recognition of tumor-associated mucin MUC1 is achieved by a fully synthetic aberrantly glycosylated MUC1 tripartite vaccine. Proc Natl Acad Sci U S A. 2012;109(1):261–6. https://doi.org/10.1073/pnas.1115166109.
65. Cai H, Chen M-S, Sun Z-Y, Zhao Y-F, Kunz H, Li Y-M. Self-adjuvanting synthetic antitumor vaccines from MUC1 Glycopeptides conjugated to T-cell epitopes from tetanus toxoid. Angew Chem Int Ed Engl. 2013;52(23):6106–10. https://doi.org/10.1002/anie.201300390.
66. Palitzsch B, Hartmann S, Stergiou N, Glaffig M, Schmitt E, Kunz H. A fully synthetic four-component antitumor vaccine consisting of a mucin Glycopeptide antigen combined with three different T-helper-cell epitopes. Angew Chem Int Ed Engl. 2014;53(51):14245–9. https://doi.org/10.1002/anie.201406843.
67. Thompson P, Lakshminarayanan V, Supekar NT, Bradley JM, Cohen PA, Wolfert MA, Gendler SJ, Boons G-J. Linear synthesis and immunological properties of a fully synthetic vaccine candidate containing a sialylated MUC1 glycopeptide. Chem Commun. 2015;51(50):10214–7. https://doi.org/10.1039/C5CC02199E.
68. Liao G, Zhou Z, Suryawanshi S, Mondal MA, Guo Z. Fully synthetic self-Adjuvanting alpha-2,9-Oligosialic acid based conjugate vaccines against group C meningitis. ACS Cent Sci. 2016;2(4):210–8. https://doi.org/10.1021/acscentsci.5b00364.
69. Lewicky JD, Ulanova M, Jiang ZH. Synthesis of a TLR4 agonist-carbohydrate antigen conjugate as A self-adjuvanting cancer vaccine. ChemistrySelect. 2016;5:906–10.

Intracellular TLR4 Signaling

Nikolay N. Kuzmich

Abstract Intracellular part of TLR4 signaling starts on the inner side of the cytoplasm and consists of two branches, namely, TLR4/TRIF/IRF3 and TLR4/MyD88/NF-κB. Through complex interactions including phosphorylation and ubiquitination, they lead to the activation of various transcription factors and are intrinsically regulated.

Keywords TLR4 · NF-κB · IRF3 · Adaptor proteins · Signal transduction

1 Introduction

Cytoplasmic section of TLR4 signal transduction is important for the transcription factor activation necessary for the host immune response and for apoptotic and other pathways as well. Transferring LPS from CD14 to MD2 induces dimerization of the TLR4-MD2 complex via leucine-rich repeat motifs outside the membrane and Toll/interleukin-1 receptor (TIR) domains [74]. On the cytoplasmic side of the membrane, the TIR-TIR surface forms a starting platform for further interactions.

Next, the signal transduction can follow one of the two possible directions of TLR4 signaling pathway, namely, TLR4/MyD88/NF-kB and TLR4/TRIF/IRF3 (Fig. 1). Among the other toll-like receptors, TLR4 is unique in that way it uses both MyD88 and TRIF pathways. On the contrary, toll-like receptors 1, 2, and 5–9 signal only via MyD88, and TRIF is used by TLR3 only [41]. For the next stages of signal transduction, the adaptor proteins MyD88 (myeloid differentiation primary response gene 88), TIRAP (TIR domain-containing adaptor protein), TRAM (TRIF-related adaptor molecule), and TRIF (TIR-domain-containing adapter-inducing interferon-β) are necessary [67, 68]. The TIRAP's alternative name is Mal, MyD88

N. N. Kuzmich (✉)
Department of Drug Safety, Smorodintsev Research Institute of Influenza, WHO National Influenza Centre of Russia, 15/17 prof. Popov str., Saint-Petersburg 197376, Russia

Laboratory of Bioinformatics, Institute of Biotechnology and Translational Medicine, I. M. Sechenov First Moscow State Medical University, 8-2 Trubetskaya str., Moscow 119991, Russia

C. Rossetti, F. Peri (eds.), *The Role of Toll-Like Receptor 4 in Infectious and Non Infectious Inflammation*, Progress in Inflammation Research 87,
https://doi.org/10.1007/978-3-030-56319-6_2

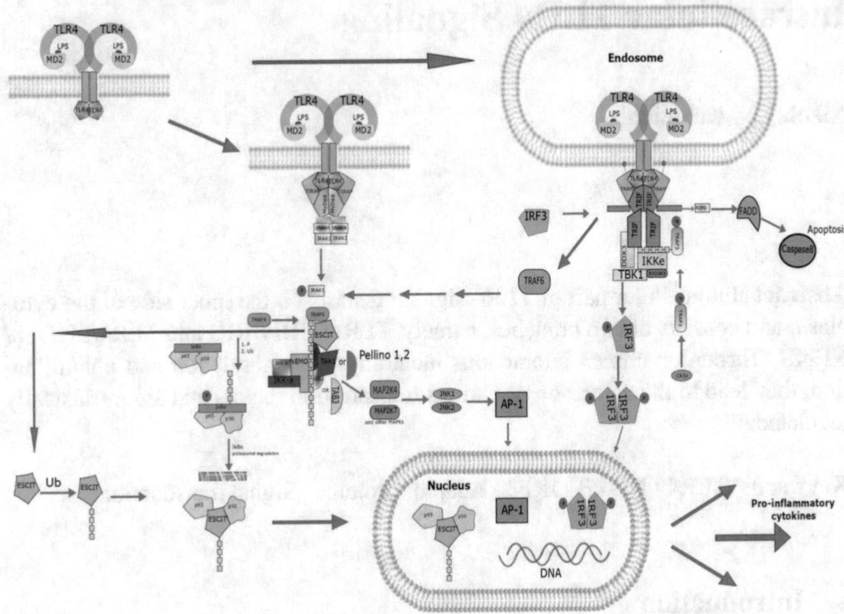

Fig. 1 Intracellular TLR4 signaling pathway. Note: the scheme is approximate and does not reflect the overall complexity as well as exact topology and stoichiometry of particular complexes and interactions

adapter-like protein. The TRIF-IRF3 (IRF3 means interferon regulatory factor 3) pathway is also used by another toll-like receptor (TLR3) but it recruits TRIF directly, without TRAM involvement [69]. The adaptor proteins are suggested to stabilize TLR4 TIR dimer [6].

Just as the intracellular part of TLR4, all the TLR4 adaptor proteins contain TIR-domain [66]. Structurally, this motif consists of central five-stranded parallel β-sheet surrounded by five α-helices.

The pathways starting from MyD88 and from TRIF adaptors compete with each other and are mutually exclusive [25]. The TLR4/MyD88 pathway begins with TLR4/MD2/LPS complex situated on plasma membrane. Meanwhile, TLR4/TRIF transduction begins after TLR4/MD2/LPS internalization in endosomes. TRAM dissociates from the membrane to endosomes after LPS stimulation, starting there the TRAM-IRF3 pathway [37, 57]. Preventing TLR4 endocytosis by LPS-stimulated cells blocks the TRIF-dependent branch of TLR4 signaling [86].

2 TLR4/MyD88/NF-kB Pathway

The signal transduction in MyD88-dependent pathway proceeds by TIRAP recruitment to the TIR-TIR part of TLR4 dimers and the binding occurs via TIRAP TIR-domain. TIRAP is located in cytoplasm and possesses a phosphatidylinositol 4,5-bisphosphate-binding domain that provides membrane anchoring and forms a homodimer in vivo [51, 94] (Fig. 2c). Both TIRAP and TRAM need to be tyrosine-phosphorylated for further activation of both pathways [23, 33, 76].

MyD88 molecules form a homo-dimer in solution meditated by their N-terminal death domains (DD) and bind to TIRAP by their own C-terminal TIR-domains. Unlike TIRAP, MyD88 itself cannot bind TLR4 TIR-domain as demonstrated by in vitro experiments [65]. Interestingly, TIR-domains of TLR4 and TIRAP are able to form poly-TIR homo- and heterofilaments in vitro [95]. TIRAP but not TLR4-TIR also induced the formation of MyD88 oligomers. Upon binding to TIRAP-TLR4 in vivo, MyD88 oligomerizes further and recruits the Interleukin-1 receptor-associated serine/threonine kinases (IRAK), namely, IRAK4 and IRAK2 or IRAK1 via the death domains (DD), forming so-called myddosome. Death domain (Fig. 2d) is a conserved interaction element that occurs by many protein families [17, 98] and consists of six alpha-helices. Nearly all DDs possess Arg-X-Asp-Leu motif at positions 78–81 [73]. The structural ensemble of MyD88-IRAK2-IRAK4 DDs has been resolved [50] by X-ray crystallography. The stoichiometric ratio of MyD88-IRAK2-IRAK4 helical tower ensemble is 6:4:4 and its approximate size is 70x110Å. The myddosome formation induces IRAK4 auto-phosphorylation [20, 97]. IRAK1 can also bind to the MyD88-IRAK4 complex and be phosphorylated by IRAK4 [3]. Sequence alignment analysis has shown that the IRAK2 residues critical for interaction with the MyD88-IRAK4 complex are very similar to those in IRAK1 [50]. A clinical study has demonstrated the predisposition to invasive bacterial disease by MyD88- and IRAK4-deficient patients [77]. This disease is defined here as a pathology due to the presence of a disease-causing bacterium in a fluid or tissue that is normally sterile. The role of MyD88 in the innate immune signaling has been reviewed by Deguine and Barton [111]. The detailed review for myddosome signaling has been published recently [3].

Once phosphorylated, IRAK1 separates from the myddosome. Next, the TNF receptor associated factor 6 (TRAF6) that forms a mushroom-like trimer via its C-domains [25] (also known as TRAF domains) is recruited. TRAF6 belongs to the group of TRAF proteins, the key adaptor molecules possessing E3 ubiquitin ligase activity [44]. The C-domain that is conserved throughout TRAF protein family binds to phosphorylated IRAK1 and then promotes poly-ubiquitination of itself within RING domain at the Lys63 [11] site. Mechanistically, E2 ligase Ubc13-Uev1A assists TRAF6 in the formation of ubiquitin chains [103]. The TRAF6 RING domain is responsible for interaction with E2-ligase/ubiquitin complex (Fig. 2b). It was demonstrated experimentally [101] that TRAF6 forms an endogenous high mass signaling complex together with ESCIT (Evolutionarily Conserved Signaling Intermediate in Toll pathways) [58] and TAK1 (transforming growth

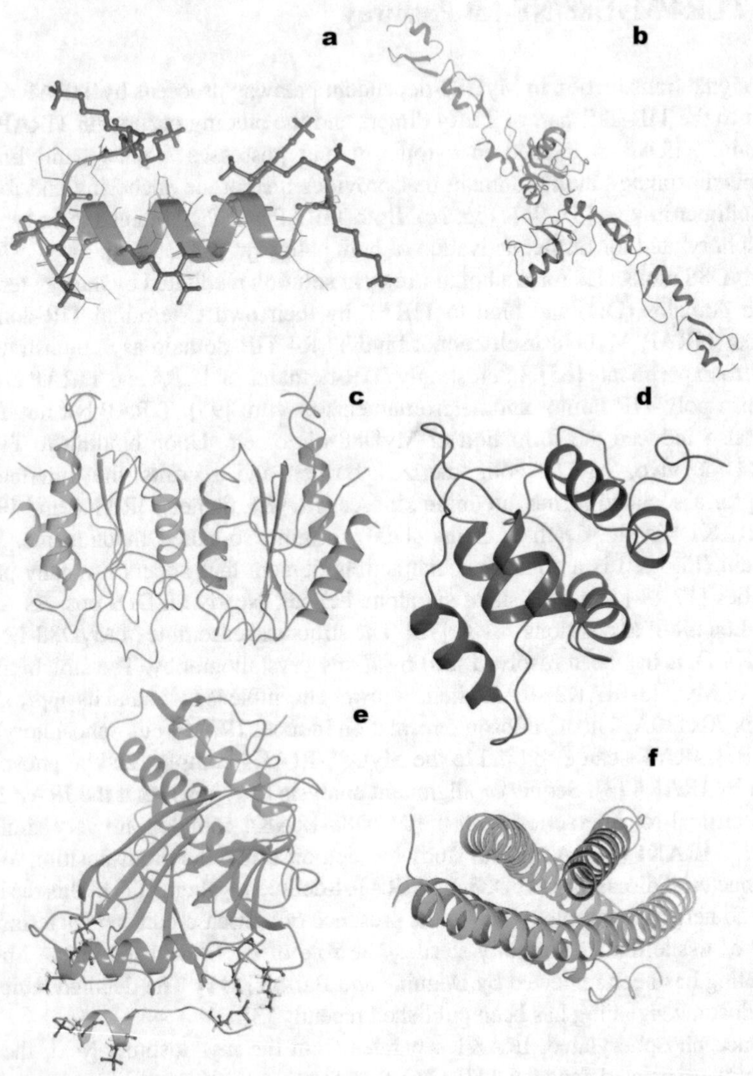

Fig. 2 Examples of structural motifs occurring in the intracellular TLR4 signaling pathway. (**a**) TIRAP phosphoinositide-binding motif (PDB: 5T7Q). Basic residues are colored azure. (**b**) TRAF6 RING domain dimer (PDB: 3HCS). (**c**) TIRAP TIR-domain dimer (PDB: 3UB2). (**d**) IRAK4 death domain (PDB: 2A9I). (**e**) IRF3 (green) and TRIF phosphorylated pL*x*IS motif (blue) (PDB: 5JEL). (**f**) NEMO (yellow-green) and IKKβ (red-orange) interacting coil-coiled motifs (PDB: 3BRV)

factor-β-activated kinase 1, also known as mitogen-activated protein kinase kinase kinase 7 (MAP3K7)) [63, 91], responding to LPS simulation. TRAF6 role in Lys63-ubiquitination can also be fulfilled by either Pellino-1 or Pellino-2 E3 ubiquitin ligases independently [89]. Pellino-1 is also phosphorylated and activated by IRAK1

[35, 88]. Poly-ubiquitin chains of TRAF6 alone are recognized by TAB1 and TAB2 or TAB3 (TAK1-binding proteins 1, 2, and 3) adaptor proteins of the TAK1/TAB1/TAB2 or TAK1/TAB1/TAB3 complexes and IKKγ (IkappaB kinase gamma, also known as NEMO, NF-κB essential modulator) subunit of IKK complex [19], respectively. TAB1, 2, and 3 co-localize with IKK complex on the Lys63 as well as linear N-end methionine (Met1) [10] ubiquitinated chains [13]. This allows recruitment and activation of TAK1 by its auto-phosphorylation and consequent phosphorylation of IκB complex, respectively [39].

TAB adaptors are also necessary for directing the TAK1-dependent activation of MAP kinase kinases to switch on JNK1/2 (c-Jun N-terminal kinases 1 and 2) and p38γ MAP kinases [47, 110]. This activation pathway was significantly diminished in TAB2/TAB3 double knockout cells although there are evidences that TAK1 activation can also go without TAB2/TAB3 [110] and along with NF-κB they induce production and release of pro-inflammatory cytokines IL-1β, TNF-α, and IL-6 [1].

The IKK complex consists of two catalytic subunits, namely, IKKα (CHUK) and IKKβ, and a non-catalytic regulatory subunit NEMO (IKKγ). IKKα and IKKβ themselves have distinct functions and substrate specificities albeit a high degree of sequence similarity [26]. The IKK binding domain of NEMO contains a coiled coil motif [4] (Fig. 2f). In the IKK complex, the NEMO unit is responsible for the interaction with poly-ubiquitinated chains.

NF-κB activation requires the phosphorylation of its inhibitor IκBα by IKK kinase. Phosphorylation occurs at sites Ser32 and Ser36, which marks IκBα for Lys48-linked poly-ubiquitination and consequent proteasomal degradation by 26S pathway [56]. It leads to release and activation of p50 and p65 elements of NF-κB. The details of this process have been described in review by Sun [90]. NF-κB has many functions including growth control [30], tumorigenesis [102], and apoptosis [15]. ESCIT forms a complex with p50 and p65 after ubiquitination at Lys372 followed by translocation to the nucleus. ESCIT participation proved to be essential in TLR4-meditated NF-κB activation [101].

Apart from the canonical NF-κB activation, IKK components (IKKα) are involved in overall transcription activation in particular by promoting histone H3 phosphorylation and binding to CREBBP (cAMP response element-binding protein) [104].

Quite recently, Sanjo et al. discovered previously unknown TAK1 capability to protect macrophages from TLR4-induced pro-inflammatory cell death [83]. The experimental data demonstrated that TAK1 is important to escape macrophage TLR4/TRIF/Caspase8-mediated cell death and to decrease septic shock severity in vivo. This underlines the TAK1 role in maintenance of the balance in host immune response.

MyD88 can also bind to TRAF3 via its C-domain and activate it leading to IFN-β production via IRF3 and IRF7 activation [64], as well as anti-inflammatory cytokine IL-10 [27].

3 TLR4/TRIF/IRF3 Branch

Analogously, binding the TRAM homodimer to the intracellular TLR4-TIR domains is necessary for adaptor recruitment in the TLR4/IRF3 pathway [14]. Just as TIRAP, TRAM is a membrane-bound bridging adaptor [22].

There are at least several factors contributing to TLR4/MD2/LPS internalization to endosomes and re-directing the signaling route to TRIF/IRF3 pathway. At the early stage, CD14 promotes the formation of endosomes containing TLR4-MD2-LPS complex [107]. Both TIRAP and TRAM adaptors possess a basic motif [36, 37] (Fig. 2a) responsible for binding to membrane phosphatidylinositol-(4,5)-bisphosphate (PtdIns(4,5)P2). In addition to this motif, TRAM also has N-myristoyl anchor to plasma membrane [81]. This anchor is important for TRAM to be active as demonstrated by site-specific mutagenesis study [81]. Many studies have shown that N-myristoylation of proteins plays a significant role in many immune cell signaling cascades [93]. PI3K kinase isoform p110δ is also involved in TLR4 internalization. It regulates the concentration of PtdIns(4,5)P2 [112] in the membrane transforming it into phosphatidylinositol-(3,4,5)-trisphosphate, and as PtdIns(4,5) P2 level drops, TIRAP is detached from the membrane, shifting the TIRAP-TRAM equilibrium in favor of the latter.

Trafficking of TRAM from the endocytic recycling compartment is controlled by glycoprotein SLAMF1 (signaling lymphocytic activation molecule family 1), also known as CD150 [106]. Silencing of SLAMF1 by siRNA strongly reduced TLR4-mediated IFNβ mRNA expression. It was also demonstrated that SLAMF1 binds by its C-terminal part to the N-terminal part of TRAM TIR-domain. TRAM was also found to be required for phagocytosis in human macrophages [87]. Raftlin (RFTN1) is also responsible for TLR4 endocytosis via interaction with clathrin–AP-2 complex [92]. Annexin A2 is recruited together with TRAM upon LPS stimulation and binds directly to TLR4. This fact was confirmed biochemically and is supported by protein–protein docking studies [109]. It results in accelerated compartmentalization of TLR4 to the early endosomes and up-regulated IRF3 activation. In the context of in vivo effects, annexin A2 attenuated bacteria-induced pulmonary inflammation and improved host-mediated intra-abdominal pathogen clearance compared to the *anxa2*$^{-/-}$ animals.

Tyrosine-protein phosphatase non-receptor type 4 (PTPN4) is responsible for the removal of the phosphate group attached to Tyr167 of TRAM, and the following inhibition of TRAM–TRIF interaction, TLR4-induced IRF3 activation, and subsequent IFN-β production [33].

TRAM also activates the TRAF6-dependent NF-κB activation, and TRAM-TRAF6 association was confirmed experimentally [96]. Interestingly, TRIF also can activate NF-κB via TRAF6, binding to its N-terminus [84].

Next, TRIF is recruited to the endosomal TRAM-(TLR4 TIR) complex binding by its TIR-domain. For TRIF to be active, it should form a homo-oligomer [21]. The importance of TRIF for LPS response have been previously demonstrated in vivo, when TRIF-KO mice were protected from severe sepsis in the cecal ligation and puncture (CLP) model [29]. The C-terminal part of TRIF comprises receptor-interacting protein (RIP) homotypic interaction motif important for FADD (Fas-associated death domain protein)-dependent apoptosis pathway [38]. Toward N-end begins TIR-domain, which is responsible for interaction with TLR4/TRAM complex.

Serine-threonine kinase TBK1 (TANK binding kinase-1) activates both NF-kB [78] and IRF3 branches [85] by phosphorylation of TRIF, which leads to recruiting IRF3 [53]. IRF3 is phosphorylated at Ser396 or other sites in the proximity [82] with participation of either TBK1 or IKKε. Serine/threonine-protein kinase RIO3K3 also serves here as an adaptor protein [18] and physically interacts with TBK1 and IRF3. Another player in this game is DEAD box protein 3 (DDX3), which interacts with both TBK1 and IKKε. It promotes IKKε auto-phosphorylation and consequently enhances phosphorylation of IRF3 [24]. TRAF3 is phosphorylated by conserved serine/threonine kinase CK1ε at Ser349 site, which leads to TRAF3 auto-ubiquitination, facilitating its complexing with TBK1 [112] and TBK1 auto-phosphorylation. The TBK1/IKKε complex interacts with TRAF3 via the scaffold dimerization domain (SDD) of TBK1 [16]. E3 ubiquitin-protein ligase homologous to the E6-AP carboxyl terminus domain-containing protein 3 (HECTD3) mediates TRAF3 poly-ubiquitination enabling TBK1-TRAF3 complex formation [49]. Structurally, TRIF and IRF3 share pL*x*IS conserved motif, which serves as a phosphorylation site. Their interaction is stabilized by ion-ionic bonds between TRIF phosphate groups and IRF3 adjacent Lys and Arg residues (Fig. 2e).

Along with TBK1, IKKε is a so-called non-canonical IKK kinase. TBK1 and IKKε kinases regulate the integrity of pathogen-containing vacuoles and restrict bacterial proliferation in the cytosol [106].

Keeping in mind the overall complexity of interactions and the number of participating signaling/adaptor components involved, one can hypothesize that all these partners form relatively large macro- or supramolecular complexes [22]. Analogously to the aforementioned myddosome, the term triffosome for this complex has been suggested. Its exact topology and stoichiometry remains unclear. Crystallographic structural data for some pairwise interactions between the complex partners are available.

IRF3 then dimerizes and translocates to the nucleus to initiate the transcription of the IFN-β gene [32]. Just as NF-κB, IRF3 also interacts with CREBBP [80].

TBK1 and IKKε can also be phosphorylated by IKKα and IKKβ kinase [8]. The results of at least two studies support the IRAK1 role as a negative regulator of IRF3 activation [7].

THO (suppressor of the Transcriptional defect of Hpr1 by Overexpression) complex subunit 7 homolog (THOC7), suppressor of cytokine signaling 3 (SOCS3), and TRAF-interacting protein (TRIP) were found to promote TBK1 proteasomal degradation and thus act as negative regulators of IRF3 activation [28, 52, 108]. Phosphatases Cdc25A and PPM1B promote TBK1 dephosphorylation and decrease the phosphorylation of its downstream substrate IRF3 [79, 111].

MyD88-signaling provides more rapid activation of TBK1 than the TRIF-dependent activation, which requires rate-limiting association with OPTN (optineurin) and TANK [2]. The role of OPTN in the interferon pathway and its interaction with TBK1 has been reviewed [70, 99].

4 Regulation of the TLR4 Intracellular Signaling

The excessive immune response (cytokine production, etc.) can be deleterious for the host organism so the regulation measures have been foreseen by the evolution.

The mechanisms impeding transduction include dephosphorylation, de-ubiquitination, proteasomal degradation, and competitive inhibition. They will be considered further in the downstream order.

At very early stage, kinase SFK (Src family of protein tyrosine kinases) makes TLR4 tyrosine-phosphorylated and promotes the dissociation of MyD88 and TIRAP from TLR4 intracellular TIR-domain [62]. As a result, LPS-promoted activation of NF-κB and JNK1/2 is not observed.

The A20 binding inhibitor of NF-κB1 (ABIN1) expression is increased early by the activation of MyD88 signaling cascade. It binds to polyubiquitin chains disrupting MAPKs and IKK activation complexes and prevents the overproduction of inflammatory mediators by NF-κB pathway [54].

IRAK-M kinase prevents the formation of IRAK-TRAF6 complexes [46]. The deficiency of this IRAK-kinase in mice exacerbated neurovascular damages induced by cerebral ischemia and reperfusion [55]. Recruitment of TRAF6 to IRAK1 is also prevented by major vault protein (MVP) as shown by in vitro studies [5]. Association of evolutionarily conserved signaling intermediate in Toll pathways (ESCIT) with TRAF6 was shown to be inhibited by p62 [45]. PRDX1 (peroxiredoxin 1) inhibited TRAF6-ubiquitin-ligase activity and NF-κB activation [61]. A similar effect has cereblon over-expression upon LPS stimulation both in vitro and in vivo [59]. TAK1 activity is negatively regulated by ribosomal S6 kinase 1 (S6K1) [43], which interferes with the interaction between TAK1 and TAB1. The zinc fingers of the terminal C-domain bind to ubiquitinated oligomers (both Lys63 and Met1) preventing their interactions with TAK1 and IKK complexes. More than that, phosphorylation by IKKβ leads to accelerated hydrolysis of Lys63-ubiquitin linkage [100]. IKK activation by TAK1 with its adaptors is inhibited by de-ubiquitination. Ubiquitin-specific protease 18 (USP18) cleaves the ubiquitin chains on NEMO and TAK1 itself [105].

Peroxiredoxin-6 (Prdx6) was also found to inhibit MyD88 – NF-κB pathway by interaction with TRAF-C-domain of TRAF6 and preventing association of TRAF6 with ECSIT [60], which is necessary for further signal transduction as mentioned above. Interaction of cereblon (CRBN), a substrate receptor protein for the CRL4A E3 ubiquitin ligase complex with zinc finger of TRAF6, leads to attenuation of TRAF6 ubiquitination, as found by Min et al. [59]. The double-knockout CRBN mice have worse survival after LPS challenge and also have increased levels of inflammatory cytokines. CRBN was also found to interact with TAK1 as well but it did not affect the signal transduction to NF-kB. Post-translational modifications of the TAK1-TAB complex are discussed in detail in the review [31]. E3 ubiquitin-protein ligase Cbl ubiquitinates TRAF6 at Lys48 site. This ubiquitination site, unlike Lys63, is NF-kB-independent and leads to TRAF6 proteosomal degradation. So c-Cbl overexpression leads to significant suppression of the transcriptional activity of NF-κB [34].

TRAF family member-associated NF-κB activator (TANK) provides negative regulation of the canonical IKKs [9], leading to attenuation of the host immune response. TANK is also responsible for prevention of autoimmune disorders [40]. The *Tank*$^{-/-}$ mice developed glomerulonephritis followed by deaths from 3 months after birth.

Targeting TBK1 by tripartite motif-containing protein 11 (TRIM11) negatively affects IFN-β production [48]. THO complex subunit 7 homolog (THOC7) also promotes proteasomal degradation of TBK1 through a ubiquitin-dependent degradation system [28]. THOC7 knockdown significantly increased IRF3 dimerization and upregulated type I IFN production.

Prostaglandin E2 (PGE2) was found to inhibit TLR4-TRIF signaling branch and restricts IRF3 activation both in vitro and in vivo [75]. Deubiquitinating enzyme A (DUBA) has been shown to inhibit the K63 ubiquitin ligase TRAF3, thus limiting IRF3 activation [42]. Regulator of calcineurin 1 (RCAN1) was found to regulate TLR4-NF-κB and TLR4-IRF3 pathways in different manners [72]. While MyD88-NF-κB-mediated cytokine levels were upregulated by RCAN1-deficient mice compared to the wild type animals, production of cytokines mediated by the TRIF-TLR4 branch was significantly suppressed. The TLR4/IRF3 branch is also suppressed by transmembrane emp24 domain-containing protein 7 (TMED7). Following LPS stimulation, TMED7 relocates from a Golgi-like perinuclear structure into the endosome and disrupts TRIF–TRAM interaction. Also, TMED7 was found to promote TLR4 degradation following LPS stimulation [12]. TAG ("TRAM adaptor with GOLD domain"), being a splice variant of TRAM, also inhibited TRIF–TRAM interaction in a competitive manner [71].

It can be concluded that almost every stage of the signaling cascade has a concurrent inhibiting mechanism impeding the activation and holding it moderate for overall balance.

5 Perspectives

The knowledge of the intracellular TLR4 signaling has greatly progressed over the two last decades. Very important advance has been achieved in the structure determination of signaling pathway components and a lot of crystallographic data became available. It shed light on the molecular mechanisms of protein–protein interactions and opened a door to in silico design of the pathway modulators, which can be useful for treatment of acute and chronic inflammatory disorders. Especially, it is worth noting the discovery of the helical myddosome DD-domain ensemble.

However, despite the successes achieved not all the details of the pathways have been clarified for today. To be exact, the participating proteins are reported every year. From one hand, the number of white spots decreases but simultaneously the new questions evolve. The failures in clinical trials probably reflect our lack of knowledge in the particular signaling mechanism that was attempted to be blocked. For instance, the bypass interactions or the regulation appears to be more complicated than expected and also the inhibition of a particular step may have the repercussions affecting other important cell mechanisms.

Nevertheless, the investigating TLR4 and adjacent signaling pathways remain to be an important area of molecular immunology.

References

1. Akira S, Takeda K. Toll-like receptor signaling. Nat Rev Immunol. 2004;4(7):499–511. https://doi.org/10.1038/nri1391.
2. Bakshi S, Taylor J, Strickson S, McCartney T, Cohen P. Identification of TBK1 complexes required for the phosphorylation of IRF3 and the production of interferon beta. Biochem J. 2017;474(7):1163–74. https://doi.org/10.1042/BCJ20160992.
3. Balka KR, De Nardo D. Understanding early TLR signaling through the Myddosome. J Leukoc Biol. 2019;105(2):339–51. https://doi.org/10.1002/JLB.MR0318-096R.
4. Barczewski AH, Ragusa MJ, Mierke DF, Pellegrini M. The IKK-binding domain of NEMO is an irregular coiled coil with a dynamic binding interface. Sci Rep. 2019;9(1):2950. https://doi.org/10.1038/s41598-019-39588-2.
5. Ben J, Jiang B, Wang D, Liu Q, Zhang Y, Qi Y, Tong X, Chen L, Liu X, Zhang Y, Zhu X, Li X, Zhang H, Bai H, Yang Q, Ma J, Wiemer EAC, Xu Y, Chen Q. Major vault protein suppresses obesity and atherosclerosis through inhibiting IKK-NF-kappaB signaling mediated inflammation. Nat Commun. 2019;10(1):1801. https://doi.org/10.1038/s41467-019-09588-x.
6. Bovijn C, Ulrichts P, De Smet AS, Catteeuw D, Beyaert R, Tavernier J, Peelman F. Identification of interaction sites for dimerization and adapter recruitment in toll/interleukin-1 receptor (TIR) domain of toll-like receptor 4. J Biol Chem. 2012;287(6):4088–98. https://doi.org/10.1074/jbc.M111.282350.
7. Bruni D, Sebastia J, Dunne S, Schroder M, Butler MP. A novel IRAK1-IKKepsilon signaling axis limits the activation of TAK1-IKKbeta downstream of TLR3. J Immunol. 2013;190(6):2844–56. https://doi.org/10.4049/jimmunol.1202042.
8. Clark K, Peggie M, Plater L, Sorcek RJ, Young ER, Madwed JB, Hough J, McIver EG, Cohen P. Novel cross-talk within the IKK family controls innate immunity. Biochem J. 2011a;434(1):93–104. https://doi.org/10.1042/BJ20101701.
9. Clark K, Takeuchi O, Akira S, Cohen P. The TRAF-associated protein TANK facilitates cross-talk within the IkappaB kinase family during toll-like receptor signaling. Proc Natl Acad Sci U S A. 2011b;108(41):17093–8. https://doi.org/10.1073/pnas.1114194108.
10. Cohen P, Strickson S. The role of hybrid ubiquitin chains in the MyD88 and other innate immune signaling pathways. Cell Death Differ. 2017;24(7):1153–9. https://doi.org/10.1038/cdd.2017.17.
11. Deng L, Wang C, Spencer E, Yang L, Braun A, You J, Slaughter C, Pickart C, Chen ZJ. Activation of the IkappaB kinase complex by TRAF6 requires a dimeric ubiquitin-conjugating enzyme complex and a unique polyubiquitin chain. Cell. 2000;103(2):351–61. https://doi.org/10.1016/s0092-8674(00)00126-4.
12. Doyle SL, Husebye H, Connolly DJ, Espevik T, O'Neill LA, McGettrick AF. The GOLD domain-containing protein TMED7 inhibits TLR4 signaling from the endosome upon LPS stimulation. Nat Commun. 2012;3:707. https://doi.org/10.1038/ncomms1706.

13. Emmerich CH, Ordureau A, Strickson S, Arthur JS, Pedrioli PG, Komander D, Cohen P. Activation of the canonical IKK complex by K63/M1-linked hybrid ubiquitin chains. Proc Natl Acad Sci U S A. 2013;110(38):15247–52. https://doi.org/10.1073/pnas.1314715110.
14. Enokizono Y, Kumeta H, Funami K, Horiuchi M, Sarmiento J, Yamashita K, Standley DM, Matsumoto M, Seya T, Inagaki F. Structures and interface mapping of the TIR domain-containing adaptor molecules involved in interferon signaling. Proc Natl Acad Sci U S A. 2013;110(49):19908–13. https://doi.org/10.1073/pnas.1222811110.
15. Fan Y, Dutta J, Gupta N, Fan G, Gelinas C. Regulation of programmed cell death by NF-kappaB and its role in tumorigenesis and therapy. Adv Exp Med Biol. 2008;615:223–50. https://doi.org/10.1007/978-1-4020-6554-5_11.
16. Fang R, Jiang Q, Zhou X, Wang C, Guan Y, Tao J, Xi J, Feng JM, Jiang Z. MAVS activates TBK1 and IKKepsilon through TRAFs in NEMO dependent and independent manner. PLoS Pathog. 2017;13(11):e1006720. https://doi.org/10.1371/journal.ppat.1006720.
17. Feinstein E, Kimchi A, Wallach D, Boldin M, Varfolomeev E. The death domain: a module shared by proteins with diverse cellular functions. Trends Biochem Sci. 1995;20(9):342–4.
18. Feng J, De Jesus PD, Su V, Han S, Gong D, Wu NC, Tian Y, Li X, Wu TT, Chanda SK, Sun R. RIOK3 is an adaptor protein required for IRF3-mediated antiviral type I interferon production. J Virol. 2014;88(14):7987–97. https://doi.org/10.1128/JVI.00643-14.
19. Ferrao R, Li J, Bergamin E, Wu H. Structural insights into the assembly of large oligomeric signalosomes in the toll-like receptor-interleukin-1 receptor superfamily. Science signaling 5 (226):re3. 2012; https://doi.org/10.1126/scisignal.2003124.
20. Ferrao R, Zhou H, Shan Y, Liu Q, Li Q, Shaw DE, Li X, Wu H. IRAK4 dimerization and trans-autophosphorylation are induced by Myddosome assembly. Mol Cell. 2014;55(6):891–903. https://doi.org/10.1016/j.molcel.2014.08.006.
21. Funami K, Sasai M, Oshiumi H, Seya T, Matsumoto M. Homo-oligomerization is essential for toll/interleukin-1 receptor domain-containing adaptor molecule-1-mediated NF-kappaB and interferon regulatory factor-3 activation. J Biol Chem. 2008;283(26):18283–91. https://doi.org/10.1074/jbc.M801013200.
22. Gay NJ, Symmons MF, Gangloff M, Bryant CE. Assembly and localization of toll-like receptor signaling complexes. Nat Rev Immunol. 2014;14(8):546–58. https://doi.org/10.1038/nri3713.
23. Gray P, Dunne A, Brikos C, Jefferies CA, Doyle SL, O'Neill LA. MyD88 adapter-like (mal) is phosphorylated by Bruton's tyrosine kinase during TLR2 and TLR4 signal transduction. J Biol Chem. 2006;281(15):10489–95. https://doi.org/10.1074/jbc.M508892200.
24. Gu L, Fullam A, Brennan R, Schroder M. Human DEAD box helicase 3 couples IkappaB kinase epsilon to interferon regulatory factor 3 activation. Mol Cell Biol. 2013;33(10):2004–15. https://doi.org/10.1128/MCB.01603-12.
25. Guven-Maiorov E, Keskin O, Gursoy A, VanWaes C, Chen Z, Tsai CJ, Nussinov R. The architecture of the TIR domain signalosome in the toll-like Receptor-4 signaling pathway. Sci Rep. 2015;5:13128. https://doi.org/10.1038/srep13128.
26. Hacker H, Karin M. Regulation and function of IKK and IKK-related kinases. Sci STKE 2006 (357):re13. 2006; https://doi.org/10.1126/stke.3572006re13.
27. Hacker H, Redecke V, Blagoev B, Kratchmarova I, Hsu LC, Wang GG, Kamps MP, Raz E, Wagner H, Hacker G, Mann M, Karin M. Specificity in toll-like receptor signaling through distinct effector functions of TRAF3 and TRAF6. Nature. 2006;439(7073):204–7. https://doi.org/10.1038/nature04369.
28. He TS, Xie T, Li J, Yang YX, Li C, Wang W, Cao L, Rao H, Ju C, Xu LG. THO complex subunit 7 homolog negatively regulates cellular antiviral response against RNA viruses by targeting TBK1. Viruses. 2019;11(2) https://doi.org/10.3390/v11020158.
29. Heipertz EL, Harper J, Walker WE. STING and TRIF contribute to mouse Sepsis, depending on severity of the disease model. Shock. 2017;47(5):621–31. https://doi.org/10.1097/SHK.0000000000000771.

30. Hinz M, Krappmann D, Eichten A, Heder A, Scheidereit C, Strauss M. NF-kappaB function in growth control: regulation of cyclin D1 expression and G0/G1-to-S-phase transition. Mol Cell Biol. 1999;19(4):2690–8. https://doi.org/10.1128/mcb.19.4.2690.
31. Hirata Y, Takahashi M, Morishita T, Noguchi T, Matsuzawa A. Post-translational modifications of the TAK1-TAB complex. Int J Mol Sci. 2017;18(1) https://doi.org/10.3390/ijms18010205.
32. Honda K, Takaoka A, Taniguchi T. Type I interferon [corrected] gene induction by the interferon regulatory factor family of transcription factors. Immunity. 2006;25(3):349–60. https://doi.org/10.1016/j.immuni.2006.08.009.
33. Huai W, Song H, Wang L, Li B, Zhao J, Han L, Gao C, Jiang G, Zhang L, Zhao W. Phosphatase PTPN4 preferentially inhibits TRIF-dependent TLR4 pathway by dephosphorylating TRAM. J Immunol. 2015;194(9):4458–65. https://doi.org/10.4049/jimmunol.1402183.
34. Jang HD, Hwang HZ, Kim HS, Lee SY. C-Cbl negatively regulates TRAF6-mediated NF-kappaB activation by promoting K48-linked polyubiquitination of TRAF6. Cell Mol Biol Lett. 2019;24:29. https://doi.org/10.1186/s11658-019-0156-y.
35. Jiang Z, Johnson HJ, Nie H, Qin J, Bird TA, Li X. Pellino 1 is required for interleukin-1 (IL-1)-mediated signaling through its interaction with the IL-1 receptor-associated kinase 4 (IRAK4)-IRAK-tumor necrosis factor receptor-associated factor 6 (TRAF6) complex. J Biol Chem. 2003;278(13):10952–6. https://doi.org/10.1074/jbc.M212112200.
36. Kagan JC, Medzhitov R. Phosphoinositide-mediated adaptor recruitment controls toll-like receptor signaling. Cell. 2006;125(5):943–55. https://doi.org/10.1016/j.cell.2006.03.047.
37. Kagan JC, Su T, Horng T, Chow A, Akira S, Medzhitov R. TRAM couples endocytosis of toll-like receptor 4 to the induction of interferon-beta. Nat Immunol. 2008;9(4):361–8. https://doi.org/10.1038/ni1569.
38. Kaiser WJ, Upton JW, Mocarski ES. Receptor-interacting protein homotypic interaction motif-dependent control of NF-kappa B activation via the DNA-dependent activator of IFN regulatory factors. J Immunol. 2008;181(9):6427–34. https://doi.org/10.4049/jimmunol.181.9.6427.
39. Kanayama A, Seth RB, Sun L, Ea CK, Hong M, Shaito A, Chiu YH, Deng L, Chen ZJ. TAB2 and TAB3 activate the NF-kappaB pathway through binding to polyubiquitin chains. Mol Cell. 2004;15(4):535–48. https://doi.org/10.1016/j.molcel.2004.08.008.
40. Kawagoe T, Takeuchi O, Takabatake Y, Kato H, Isaka Y, Tsujimura T, Akira S. TANK is a negative regulator of toll-like receptor signaling and is critical for the prevention of autoimmune nephritis. Nat Immunol. 2009;10(9):965–72. https://doi.org/10.1038/ni.1771.
41. Kawai T, Akira S. The role of pattern-recognition receptors in innate immunity: update on toll-like receptors. Nat Immunol. 2010;11(5):373–84. https://doi.org/10.1038/ni.1863.
42. Kayagaki N, Phung Q, Chan S, Chaudhari R, Quan C, O'Rourke KM, Eby M, Pietras E, Cheng G, Bazan JF, Zhang Z, Arnott D, Dixit VM. DUBA: a deubiquitinase that regulates type I interferon production. Science. 2007;318(5856):1628–32. https://doi.org/10.1126/science.1145918.
43. Kim SY, Baik KH, Baek KH, Chah KH, Kim KA, Moon G, Jung E, Kim ST, Shim JH, Greenblatt MB, Chun E, Lee KY. S6K1 negatively regulates TAK1 activity in the toll-like receptor signaling pathway. Mol Cell Biol. 2014;34(3):510–21. https://doi.org/10.1128/MCB.01225-13.
44. Kim CM, Choi JY, Bhat EA, Jeong JH, Son YJ, Kim S, Park HH. Crystal structure of TRAF1 TRAF domain and its implications in the TRAF1-mediated intracellular signaling pathway. Sci Rep. 2016;6:25526. https://doi.org/10.1038/srep25526.
45. Kim MJ, Min Y, Kwon J, Son J, Im JS, Shin J, Lee KY. p62 negatively regulates TLR4 signaling via functional regulation of the TRAF6-ECSIT complex. Immune Netw. 2019;19(3):e16. https://doi.org/10.4110/in.2019.19.e16.
46. Kobayashi K, Hernandez LD, Galan JE, Janeway CA Jr, Medzhitov R, Flavell RA. IRAK-M is a negative regulator of toll-like receptor signaling. Cell. 2002;110(2):191–202. https://doi.org/10.1016/s0092-8674(02)00827-9.

47. Landstrom M. The TAK1-TRAF6 signaling pathway. Int J Biochem Cell Biol. 2010;42(5):585–9. https://doi.org/10.1016/j.biocel.2009.12.023.
48. Lee Y, Song B, Park C, Kwon KS. TRIM11 negatively regulates IFNbeta production and antiviral activity by targeting TBK1. PLoS One. 2013;8(5):e63255. https://doi.org/10.1371/journal.pone.0063255.
49. Li F, Li Y, Liang H, Xu T, Kong Y, Huang M, Xiao J, Chen X, Xia H, Wu Y, Zhou Z, Guo X, Hu C, Yang C, Cheng X, Chen C, Qi X. HECTD3 mediates TRAF3 polyubiquitination and type I interferon induction during bacterial infection. J Clin Invest. 2018;128(9):4148–62. https://doi.org/10.1172/JCI120406.
50. Lin SC, Lo YC, Wu H. Helical assembly in the MyD88-IRAK4-IRAK2 complex in TLR/IL-1R signaling. Nature. 2010;465(7300):885–90. https://doi.org/10.1038/nature09121.
51. Lin Z, Lu J, Zhou W, Shen Y. Structural insights into TIR domain specificity of the bridging adaptor mal in TLR4 signaling. PLoS One. 2012;7(4):e34202. https://doi.org/10.1371/journal.pone.0034202.
52. Liu D, Sheng C, Gao S, Yao C, Li J, Jiang W, Chen H, Wu J, Pan C, Chen S, Huang W. SOCS3 drives proteasomal degradation of TBK1 and negatively regulates antiviral innate immunity. Mol Cell Biol. 2015a;35(14):2400–13. https://doi.org/10.1128/MCB.00090-15.
53. Liu S, Cai X, Wu J, Cong Q, Chen X, Li T, Du F, Ren J, Wu YT, Grishin NV, Chen ZJ. Phosphorylation of innate immune adaptor proteins MAVS, STING, and TRIF induces IRF3 activation. Science 347 (6227):aaa2630. 2015b; https://doi.org/10.1126/science.aaa2630.
54. Lork M, Verhelst K, Beyaert R. CYLD, A20 and OTULIN deubiquitinases in NF-kappaB signaling and cell death: so similar, yet so different. Cell Death Differ. 2017;24(7):1172–83. https://doi.org/10.1038/cdd.2017.46.
55. Lyu C, Zhang Y, Gu M, Huang Y, Liu G, Wang C, Li M, Chen S, Pan S, Gu Y. IRAK-M deficiency exacerbates ischemic neurovascular injuries in experimental stroke mice. Front Cell Neurosci. 2018;12:504. https://doi.org/10.3389/fncel.2018.00504.
56. Maniatis T. A ubiquitin ligase complex essential for the NF-kappaB, Wnt/wingless, and hedgehog signaling pathways. Genes Dev. 1999;13(5):505–10. https://doi.org/10.1101/gad.13.5.505.
57. McGettrick AF, Brint EK, Palsson-McDermott EM, Rowe DC, Golenbock DT, Gay NJ, Fitzgerald KA, O'Neill LA. Trif-related adapter molecule is phosphorylated by PKC{epsilon} during toll-like receptor 4 signaling. Proc Natl Acad Sci U S A. 2006;103(24):9196–201. https://doi.org/10.1073/pnas.0600462103.
58. Mi Wi S, Park J, Shim JH, Chun E, Lee KY. Ubiquitination of ECSIT is crucial for the activation of p65/p50 NF-kappaBs in toll-like receptor 4 signaling. Mol Biol Cell. 2015;26(1):151–60. https://doi.org/10.1091/mbc.E14-08-1277.
59. Min Y, Wi SM, Kang JA, Yang T, Park CS, Park SG, Chung S, Shim JH, Chun E, Lee KY. Cereblon negatively regulates TLR4 signaling through the attenuation of ubiquitination of TRAF6. Cell Death Dis. 2016;7(7):e2313. https://doi.org/10.1038/cddis.2016.226.
60. Min Y, Wi SM, Shin D, Chun E, Lee KY. Peroxiredoxin-6 negatively regulates bactericidal activity and NF-kappaB activity by interrupting TRAF6-ECSIT complex. Front Cell Infect Microbiol. 2017;7:94. https://doi.org/10.3389/fcimb.2017.00094.
61. Min Y, Kim MJ, Lee S, Chun E, Lee KY. Inhibition of TRAF6 ubiquitin-ligase activity by PRDX1 leads to inhibition of NFKB activation and autophagy activation. Autophagy. 2018;14(8):1347–58. https://doi.org/10.1080/15548627.2018.1474995.
62. Mitchell J, Kim SJ, Seelmann A, Veit B, Shepard B, Im E, Rhee SH. Src family kinase tyrosine phosphorylates toll-like receptor 4 to dissociate MyD88 and mal/Tirap, suppressing LPS-induced inflammatory responses. Biochem Pharmacol. 2018;147:119–27. https://doi.org/10.1016/j.bcp.2017.11.015.
63. Ninomiya-Tsuji J, Kishimoto K, Hiyama A, Inoue J, Cao Z, Matsumoto K. The kinase TAK1 can activate the NIK-I kappaB as well as the MAP kinase cascade in the IL-1 signaling pathway. Nature. 1999;398(6724):252–6. https://doi.org/10.1038/18465.

64. Oganesyan G, Saha SK, Guo B, He JQ, Shahangian A, Zarnegar B, Perry A, Cheng G. Critical role of TRAF3 in the toll-like receptor-dependent and -independent antiviral response. Nature. 2006;439(7073):208–11. https://doi.org/10.1038/nature04374.
65. Ohnishi H, Tochio H, Kato Z, Orii KE, Li A, Kimura T, Hiroaki H, Kondo N, Shirakawa M. Structural basis for the multiple interactions of the MyD88 TIR domain in TLR4 signaling. Proc Natl Acad Sci U S A. 2009;106(25):10260–5. https://doi.org/10.1073/pnas.0812956106.
66. O'Neill LA, Bowie AG. The family of five: TIR-domain-containing adaptors in toll-like receptor signaling. Nat Rev Immunol. 2007;7(5):353–64. https://doi.org/10.1038/nri2079.
67. O'Neill LA, Dunne A, Edjeback M, Gray P, Jefferies C, Wietek C. Mal and MyD88: adapter proteins involved in signal transduction by toll-like receptors. J Endotoxin Res. 2003a;9(1):55–9. https://doi.org/10.1179/096805103125001351.
68. O'Neill LA, Fitzgerald KA, Bowie AG. The toll-IL-1 receptor adaptor family grows to five members. Trends Immunol. 2003b;24(6):286–90.
69. Oshiumi H, Matsumoto M, Funami K, Akazawa T, Seya T. TICAM-1, an adaptor molecule that participates in toll-like receptor 3-mediated interferon-beta induction. Nat Immunol. 2003;4(2):161–7. https://doi.org/10.1038/ni886.
70. Outlioua A, Pourcelot M, Arnoult D. The role of Optineurin in antiviral type I interferon production. Front Immunol. 2018;9:853. https://doi.org/10.3389/fimmu.2018.00853.
71. Palsson-McDermott EM, Doyle SL, McGettrick AF, Hardy M, Husebye H, Banahan K, Gong M, Golenbock D, Espevik T, O'Neill LA. TAG, a splice variant of the adaptor TRAM, negatively regulates the adaptor MyD88-independent TLR4 pathway. Nat Immunol. 2009;10(6):579–86. https://doi.org/10.1038/ni.1727.
72. Pang Z, Junkins RD, Raudonis R, MacNeil AJ, McCormick C, Cheng Z, Lin TJ. Regulator of calcineurin 1 differentially regulates TLR-dependent MyD88 and TRIF signaling pathways. PLoS One. 2018;13(5):e0197491. https://doi.org/10.1371/journal.pone.0197491.
73. Park HH, Lo YC, Lin SC, Wang L, Yang JK, Wu H. The death domain superfamily in intracellular signaling of apoptosis and inflammation. Annu Rev Immunol. 2007;25:561–86. https://doi.org/10.1146/annurev.immunol.25.022106.141656.
74. Park BS, Song DH, Kim HM, Choi BS, Lee H, Lee JO. The structural basis of lipopolysaccharide recognition by the TLR4-MD-2 complex. Nature. 2009;458(7242):1191–5. https://doi.org/10.1038/nature07830.
75. Perkins DJ, Richard K, Hansen AM, Lai W, Nallar S, Koller B, Vogel SN. Autocrine-paracrine prostaglandin E2 signaling restricts TLR4 internalization and TRIF signaling. Nat Immunol. 2018;19(12):1309–18. https://doi.org/10.1038/s41590-018-0243-7.
76. Piao W, Song C, Chen H, Wahl LM, Fitzgerald KA, O'Neill LA, Medvedev AE. Tyrosine phosphorylation of MyD88 adapter-like (mal) is critical for signal transduction and blocked in endotoxin tolerance. J Biol Chem. 2008;283(6):3109–19. https://doi.org/10.1074/jbc.M707400200.
77. Picard C, von Bernuth H, Ghandil P, Chrabieh M, Levy O, Arkwright PD, McDonald D, Geha RS, Takada H, Krause JC, Creech CB, Ku CL, Ehl S, Marodi L, Al-Muhsen S, Al-Hajjar S, Al-Ghonaium A, Day-Good NK, Holland SM, Gallin JI, Chapel H, Speert DP, Rodriguez-Gallego C, Colino E, Garty BZ, Roifman C, Hara T, Yoshikawa H, Nonoyama S, Domachowske J, Issekutz AC, Tang M, Smart J, Zitnik SE, Hoarau C, Kumararatne DS, Thrasher AJ, Davies EG, Bethune C, Sirvent N, de Ricaud D, Camcioglu Y, Vasconcelos J, Guedes M, Vitor AB, Rodrigo C, Almazan F, Mendez M, Arostegui JI, Alsina L, Fortuny C, Reichenbach J, Verbsky JW, Bossuyt X, Doffinger R, Abel L, Puel A, Casanova JL. Clinical features and outcome of patients with IRAK-4 and MyD88 deficiency. Medicine (Baltimore). 2010;89(6):403–25. https://doi.org/10.1097/MD.0b013e3181fd8ec3.
78. Pomerantz JL, Baltimore D. NF-kappaB activation by a signaling complex containing TRAF2, TANK and TBK1, a novel IKK-related kinase. EMBO J. 1999;18(23):6694–704. https://doi.org/10.1093/emboj/18.23.6694.

79. Qi D, Hu L, Jiao T, Zhang T, Tong X, Ye X. Phosphatase Cdc25A negatively regulates the antiviral immune response by inhibiting TBK1 activity. J Virol. 2018;92(19) https://doi.org/10.1128/JVI.01118-18.
80. Qin BY, Liu C, Srinath H, Lam SS, Correia JJ, Derynck R, Lin K. Crystal structure of IRF-3 in complex with CBP. Structure. 2005;13(9):1269–77. https://doi.org/10.1016/j.str.2005.06.011.
81. Rowe DC, McGettrick AF, Latz E, Monks BG, Gay NJ, Yamamoto M, Akira S, O'Neill LA, Fitzgerald KA, Golenbock DT. The myristoylation of TRIF-related adaptor molecule is essential for toll-like receptor 4 signal transduction. Proc Natl Acad Sci U S A. 2006;103(16):6299–304. https://doi.org/10.1073/pnas.0510041103.
82. Sabokbar A, Afrough S, Mahoney DJ, Uchihara Y, Swales C, Athanasou NA. Role of LIGHT in the pathogenesis of joint destruction in rheumatoid arthritis. World J Exp Med. 2017;7(2):49–57. https://doi.org/10.5493/wjem.v7.i2.49.
83. Sanjo H, Nakayama J, Yoshizawa T, Fehling HJ, Akira S, Taki S. Cutting edge: TAK1 safeguards macrophages against Proinflammatory cell death. J Immunol. 2019;203(4):783–8. https://doi.org/10.4049/jimmunol.1900202.
84. Sato S, Sugiyama M, Yamamoto M, Watanabe Y, Kawai T, Takeda K, Akira S. Toll/IL-1 receptor domain-containing adaptor inducing IFN-beta (TRIF) associates with TNF receptor-associated factor 6 and TANK-binding kinase 1, and activates two distinct transcription factors, NF-kappa B and IFN-regulatory factor-3, in the toll-like receptor signaling. J Immunol. 2003;171(8):4304–10. https://doi.org/10.4049/jimmunol.171.8.4304.
85. Sharma S, tenOever BR, Grandvaux N, Zhou GP, Lin R, Hiscott J. Triggering the interferon antiviral response through an IKK-related pathway. Science. 2003;300(5622):1148–51. https://doi.org/10.1126/science.1081315.
86. Shim DW, Heo KH, Kim YK, Sim EJ, Kang TB, Choi JW, Sim DW, Cheong SH, Lee SH, Bang JK, Won HS, Lee KH. Anti-inflammatory action of an antimicrobial model peptide that suppresses the TRIF-dependent signaling pathway via inhibition of toll-like receptor 4 endocytosis in lipopolysaccharide-stimulated macrophages. PLoS One. 2015;10(5):e0126871. https://doi.org/10.1371/journal.pone.0126871.
87. Skjesol A, Yurchenko M, Bosl K, Gravastrand C, Nilsen KE, Grovdal LM, Agliano F, Patane F, Lentini G, Kim H, Teti G, Kumar Sharma A, Kandasamy RK, Sporsheim B, Starheim KK, Golenbock DT, Stenmark H, McCaffrey M, Espevik T, Husebye H. The TLR4 adaptor TRAM controls the phagocytosis of gram-negative bacteria by interacting with the Rab11-family interacting protein 2. PLoS Pathog. 2019;15(3):e1007684. https://doi.org/10.1371/journal.ppat.1007684.
88. Smith H, Peggie M, Campbell DG, Vandermoere F, Carrick E, Cohen P. Identification of the phosphorylation sites on the E3 ubiquitin ligase Pellino that are critical for activation by IRAK1 and IRAK4. Proc Natl Acad Sci U S A. 2009;106(12):4584–90. https://doi.org/10.1073/pnas.0900774106.
89. Strickson S, Emmerich CH, Goh ETH, Zhang J, Kelsall IR, Macartney T, Hastie CJ, Knebel A, Peggie M, Marchesi F, Arthur JSC, Cohen P. Roles of the TRAF6 and Pellino E3 ligases in MyD88 and RANKL signaling. Proc Natl Acad Sci U S A. 2017;114(17):E3481–9. https://doi.org/10.1073/pnas.1702367114.
90. Sun SC. The non-canonical NF-kappaB pathway in immunity and inflammation. Nat Rev Immunol. 2017;17(9):545–58. https://doi.org/10.1038/nri.2017.52.
91. Takaesu G, Kishida S, Hiyama A, Yamaguchi K, Shibuya H, Irie K, Ninomiya-Tsuji J, Matsumoto K. TAB2, a novel adaptor protein, mediates activation of TAK1 MAPKKK by linking TAK1 to TRAF6 in the IL-1 signal transduction pathway. Mol Cell. 2000;5(4):649–58. https://doi.org/10.1016/s1097-2765(00)80244-0.
92. Tatematsu M, Yoshida R, Morioka Y, Ishii N, Funami K, Watanabe A, Saeki K, Seya T, Matsumoto M. Raftlin controls lipopolysaccharide-induced TLR4 internalization and TICAM-1 signaling in a cell type-specific manner. J Immunol. 2016;196(9):3865–76. https://doi.org/10.4049/jimmunol.1501734.

93. Udenwobele DI, Su RC, Good SV, Ball TB, Varma Shrivastav S, Shrivastav A. Myristoylation: An important protein modification in the immune response. Front Immunol. 2017;8:751. https://doi.org/10.3389/fimmu.2017.00751.
94. Valkov E, Stamp A, Dimaio F, Baker D, Verstak B, Roversi P, Kellie S, Sweet MJ, Mansell A, Gay NJ, Martin JL, Kobe B. Crystal structure of toll-like receptor adaptor MAL/TIRAP reveals the molecular basis for signal transduction and disease protection. Proc Natl Acad Sci U S A. 2011;108(36):14879–84. https://doi.org/10.1073/pnas.1104780108.
95. Ve T, Vajjhala PR, Hedger A, Croll T, DiMaio F, Horsefield S, Yu X, Lavrencic P, Hassan Z, Morgan GP, Mansell A, Mobli M, O'Carroll A, Chauvin B, Gambin Y, Sierecki E, Landsberg MJ, Stacey KJ, Egelman EH, Kobe B. Structural basis of TIR-domain-assembly formation in MAL- and MyD88-dependent TLR4 signaling. Nat Struct Mol Biol. 2017;24(9):743–51. https://doi.org/10.1038/nsmb.3444.
96. Verstak B, Stack J, Ve T, Mangan M, Hjerrild K, Jeon J, Stahl R, Latz E, Gay N, Kobe B, Bowie AG, Mansell A. The TLR signaling adaptor TRAM interacts with TRAF6 to mediate activation of the inflammatory response by TLR4. J Leukoc Biol. 2014;96(3):427–36. https://doi.org/10.1189/jlb.2A0913-487R.
97. Vollmer S, Strickson S, Zhang T, Gray N, Lee KL, Rao VR, Cohen P. The mechanism of activation of IRAK1 and IRAK4 by interleukin-1 and toll-like receptor agonists. Biochem J. 2017;474(12):2027–38. https://doi.org/10.1042/BCJ20170097.
98. Weber CH, Vincenz C. The death domain superfamily: a tale of two interfaces? Trends Biochem Sci. 2001;26(8):475–81.
99. Weil R, Laplantine E, Genin P. Regulation of TBK1 activity by Optineurin contributes to cell cycle-dependent expression of the interferon pathway. Cytokine Growth Factor Rev. 2016;29:23–33. https://doi.org/10.1016/j.cytogfr.2016.03.001.
100. Wertz IE, Newton K, Seshasayee D, Kusam S, Lam C, Zhang J, Popovych N, Helgason E, Schoeffler A, Jeet S, Ramamoorthi N, Kategaya L, Newman RJ, Horikawa K, Dugger D, Sandoval W, Mukund S, Zindal A, Martin F, Quan C, Tom J, Fairbrother WJ, Townsend M, Warming S, DeVoss J, Liu J, Dueber E, Caplazi P, Lee WP, Goodnow CC, Balazs M, Yu K, Kolumam G, Dixit VM. Phosphorylation and linear ubiquitin direct A20 inhibition of inflammation. Nature. 2015;528(7582):370–5. https://doi.org/10.1038/nature16165.
101. Wi SM, Moon G, Kim J, Kim ST, Shim JH, Chun E, Lee KY. TAK1-ECSIT-TRAF6 complex plays a key role in the TLR4 signal to activate NF-kappaB. J Biol Chem. 2014;289(51):35205–14. https://doi.org/10.1074/jbc.M114.597187.
102. Xia Y, Shen S, Verma IM. NF-kappaB, an active player in human cancers. Cancer Immunol Res. 2014;2(9):823–30. https://doi.org/10.1158/2326-6066.CIR-14-0112.
103. Xie P. TRAF molecules in cell signaling and in human diseases. J Mol Signal. 2013;8(1):7. https://doi.org/10.1186/1750-2187-8-7.
104. Yamamoto Y, Verma UN, Prajapati S, Kwak YT, Gaynor RB. Histone H3 phosphorylation by IKK-alpha is critical for cytokine-induced gene expression. Nature. 2003;423(6940):655–9. https://doi.org/10.1038/nature01576.
105. Yang Z, Xian H, Hu J, Tian S, Qin Y, Wang RF, Cui J. USP18 negatively regulates NF-kappaB signaling by targeting TAK1 and NEMO for deubiquitination through distinct mechanisms. Sci Rep. 2015;5:12738. https://doi.org/10.1038/srep12738.
106. Yurchenko M, Skjesol A, Ryan L, Richard GM, Kandasamy RK, Wang N, Terhorst C, Husebye H, Espevik T. SLAMF1 is required for TLR4-mediated TRAM-TRIF-dependent signaling in human macrophages. J Cell Biol. 2018;217(4):1411–29. https://doi.org/10.1083/jcb.201707027.
107. Zanoni I, Ostuni R, Marek LR, Barresi S, Barbalat R, Barton GM, Granucci F, Kagan JC. CD14 controls the LPS-induced endocytosis of toll-like receptor 4. Cell. 2011;147(4):868–80. https://doi.org/10.1016/j.cell.2011.09.051.
108. Zhang M, Wang L, Zhao X, Zhao K, Meng H, Zhao W, Gao C. TRAF-interacting protein (TRIP) negatively regulates IFN-beta production and antiviral response by promoting

proteasomal degradation of TANK-binding kinase 1. J Exp Med. 2012;209(10):1703–11. https://doi.org/10.1084/jem.20120024.
109. Zhang S, Yu M, Guo Q, Li R, Li G, Tan S, Li X, Wei Y, Wu M. Annexin A2 binds to endosomes and negatively regulates TLR4-triggered inflammatory responses via the TRAM-TRIF pathway. Sci Rep. 2015;5:15859. https://doi.org/10.1038/srep15859.
110. Zhang J, Macartney T, Peggie M, Cohen P. Interleukin-1 and TRAF6-dependent activation of TAK1 in the absence of TAB2 and TAB3. Biochem J. 2017;474(13):2235–48. https://doi.org/10.1042/BCJ20170288.
111. Zhao Y, Liang L, Fan Y, Sun S, An L, Shi Z, Cheng J, Jia W, Sun W, Mori-Akiyama Y, Zhang H, Fu S, Yang J. PPM1B negatively regulates antiviral response via dephosphorylating TBK1. Cell Signal. 2012;24(11):2197–204. https://doi.org/10.1016/j.cellsig.2012.06.017.
112. Zhou Y, He C, Yan D, Liu F, Liu H, Chen J, Cao T, Zuo M, Wang P, Ge Y, Lu H, Tong Q, Qin C, Deng Y, Ge B. The kinase CK1varepsilon controls the antiviral immune response by phosphorylating the signaling adaptor TRAF3. Nat Immunol. 2016;17(4):397–405. https://doi.org/10.1038/ni.3395.

TLR4 Ligands: Single Molecules and Aggregates

Andra B. Schromm and Klaus Brandenburg

Abstract Lipopolysaccharide (LPS, endotoxin) is an amphipathic glycolipid that undergoes self-aggregation. The physical state and 3D organization of LPS in the aggregated state has a high impact on the biological activity and pathophysiology. Here, the basis of aggregate formation and the role of aggregate properties are presented for bacterial LPS, LPS-mimetic, and TLR4-modulating compounds with a focus on the concept of the "endotoxic conformation". A network of sequentially interacting molecules is operative to enable a sensitive and targeted delivery of LPS from aggregates to the TLR4 receptor. The structural and thermodynamic aspects of the transport and the molecular recognition of LPS by TLR4/MD-2 are presented to provide a mechanistic understanding of TLR4 activation by its ligand. Furthermore, delivery mechanisms and activation of the cytoplasmic LPS receptors caspase-4/5/11 are discussed. These insights are important for the development of new classes of immune-modulating compounds by chemical synthesis and also for modern in silico approaches to identify new lead structures for the development of therapeutics.

Keywords LPS physico-chemistry · Supramolecular structure · Lipid transport · TLR4 activation · Caspase-4/5/11 activation

1 Basis of LPS Pathophysiology

Activation of the toll-like receptor 4 (TLR4) by bacterial lipopolysaccharide (LPS, endotoxin) is among the most sensitive responses of the human immune system. LPS is the main component of the outer membrane of Gram-negative bacteria and

---

A. B. Schromm (✉)
Division of Immunobiophysics, Research Center Borstel, Leibniz Lung Center, Borstel, Germany
e-mail: aschromm@fz-borstel.de

K. Brandenburg
Brandenburg-Antiinfektiva GmbH c/o Research Center Borstel, Leibniz Lung Center, Borstel, Germany

C. Rossetti, F. Peri (eds.), *The Role of Toll-Like Receptor 4 in Infectious and Non Infectious Inflammation*, Progress in Inflammation Research 87,
https://doi.org/10.1007/978-3-030-56319-6_3

represents a central molecular trigger for the immunological recognition of an infection, the induction of inflammation, and the initiation of an antimicrobial immune response. Due to the complex physico-chemical nature of LPS, this recognition process is organized by a network of sequentially interacting molecules that have evolved to enable a sensitive and targeted delivery of LPS to cellular receptor systems. The peculiar physico-chemical behavior of LPS released from the cell wall of bacteria into aqueous environment and body fluids, the structural prerequisites for biological activity of LPS, and thermodynamic aspects of the process of molecular recognition are presented here.

The glycolipid LPS is a membrane component present exclusively in the cell envelope of Gram-negative bacteria. The cell wall of Gram-negative bacteria is organized in several layers, an inner cytoplasmic membrane, a thin peptidoglycan layer, and an outer membrane. The cytoplasmic membrane is composed of the phospholipids phosphatidylethanolamine, phosphatidylglycerol, and cardiolipin. The outer membrane contains an inner leaflet solely of phospholipids, whereas the outer leaflet is composed of LPS. Thus, LPS is the major molecule presented on the microbial surface. Chemically, LPS is composed of a bis-phosphorylated diglucosamine backbone, which is acylated in amide- and ester-linkage with up to seven fatty acids. This amphiphilic part of the molecule, termed lipid A, is the membrane anchor of LPS. In rough mutant strains, lipid A is substituted with a head group of the unusual sugar 2-keto-3-deoxyoctonate (Kdo) and further sugar residues that are distinguished as an inner and an outer core region depending on the length of the sugar substitution. In wild-type strains, an O-specific chain is attached to the outer core composed of a large number of repeating units of additional sugars [1]. The lipid A part is responsible for the immunological recognition of LPS and is thus also termed the "*endotoxic principle*" of LPS [2]. Activation of the immune system by LPS is extremely potent. LPS concentrations in the range of picogram per ml are sufficient to induce activation of TLR4 in monocytes and macrophages. The downstream signaling cascades lead to activation of nuclear-factor-κB (NF-κB) or interferon-regulatory factor 3 (IRF3) and IRF7 responsive pro-inflammatory genes [3–5]. Among LPS-TLR4 related diseases, the most harmful ones are sepsis and septic shock, pathological conditions that are accompanied by a high rate of morbidity and mortality. Excessive activation of monocytes and macrophages by bacterial pathogens leads to a dysregulated immune response. Systemic overproduction of inflammatory cytokines such as interleukin-1 (IL-1), IL-6, tumor-necrosis-factor-α (TNF-α), chemokines, and lipid mediators initiate a cascade that culminates in life-threatening organ-dysfunction and death [6, 7]. Recent data derived from single cell RNA profiling of a sepsis patient cohort indicate that during bacterial sepsis a unique immune cell signature is generated in mononuclear cells that can be clearly distinguished from other disease entities [8].

The molecular recognition of LPS by TLR4 requires its release from the bacterial cell surface. Upon cell division and cell death, endotoxin is naturally shed from the bacterial cell wall. Bacterial killing by antimicrobial immune responses of the host, mediated, for example, by complement or antimicrobial effector molecules, will liberate endotoxin from the cell wall and release it into the circulation. Of note, also the antimicrobial activity of antibiotics can lead to massive release of endotoxin into the bloodstream resulting in exaggerated immune responses that may contribute to the development of sepsis [9–12]. Various experimental animal models have demonstrated that injection of pure LPS into the blood stream or the peritoneum is sufficient to induce sepsis [13]. Galactosamine-sensitized mice represent a well-established and widely used model for the investigation of endotoxemia-associated pathology that requires low amounts of LPS in the range of 0.05–0.01 μg per animal to induce lethal effects [14]. Challenge of human volunteers with highly purified LPS is a clinical model allowing to perform highly controlled studies to investigate LPS-induced systemic inflammation in vivo [15]. These experimental model systems all demonstrate severe pathophysiological effects of purified LPS in vivo.

2 Physico-Chemistry of the TLR4 Ligand LPS

To understand the biology of LPS, its complex physico-chemistry has to be considered. LPS is an amphipathic glycolipid that similar to phospholipids undergoes thermodynamically driven self-aggregation to reduce the contact of the hydrophobic acyl chains with water. Thus, purified LPS and LPS released from the bacterial cell wall will spontaneously form aggregates to minimize the Gibbs-free energy [16]. The concentration at which molecular aggregation starts is termed the critical micellar concentration (CMC). Above the CMC, with increasing LPS concentration, the monomer concentration remains constant or is even reduced in the case of negatively charged amphiphiles, and additional molecules are incorporated into the aggregated form [17] (Fig. 1a, b). Published values of the CMC for LPS and lipid A are, however, contradictory and span a wide range of concentrations [18–22]. This may be at least partially due to limitations of the methodological approaches used. For review see [23]. Evaluating the published data, the value of the CMC for lipid A can be approximated in the concentration range $< 10^{-9}$ M.

The physical state and 3D organization of biological lipid aggregates is determined by the molecular conformation of the aggregate forming molecules. Geometric models of lipid aggregation allow an estimation of the aggregate structure based on the shape parameter $S = v / (a_0 \cdot l_c) = a_h / a_0$ (v = volume of the hydrophobic moiety, l_c length of the fully extended hydrophobic moiety, a_0, a_h cross-sectional areas of the hydrophilic and hydrophobic moiety, respectively) introduced by Israelachvili [17]. For $S < ½$ micellar structures are adopted, and in particular cases an H_I phase can also be formed. Between $S = ½$ and 1, unilamellar and multilamellar bilayer structures are favored. Whether a particular glycolipid adopts a uni- or a multilamellar structure is a complex problem, which depends, among others, on geometrical constrains, the presence of charges in the head group, the kind of counter ions, and the hydration properties of the glycolipid [16]. For $S > 1$, inverted structures such as inverted hexagonal (H_{II}) or cubic (Q) structures are formed in which the acyl chains are directed outward, and the hydrophilic moiety inward [24, 25]. In the range around $S = 1$, various phases may coexist and phase transition may be induced by

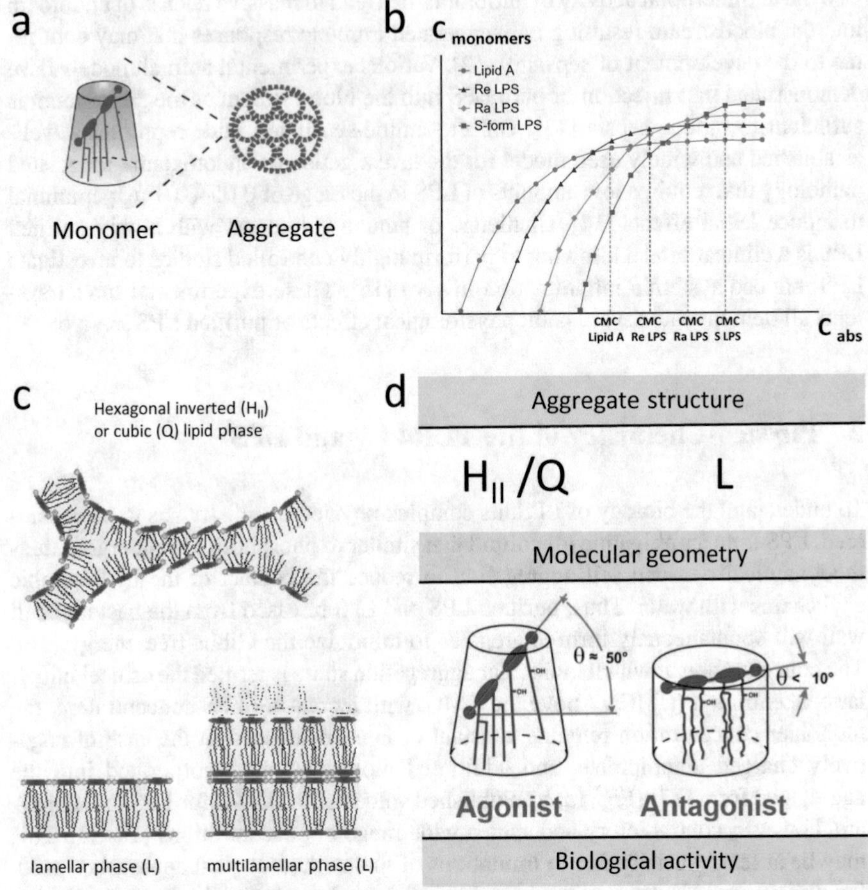

Fig. 1 Aggregation behavior, lipid phases, and molecular conformation of the TLR4 ligand LPS. (**a**) In aqueous environment, LPS undergoes aggregation to form supramolecular assemblies. (**b**) Dependence of the threshold concentration of LPS aggregate formation (CMC, critical micellar concentration) on the carbohydrate content of LPS. Theoretical concentration monomers versus absolute concentration C_{abs} for different endotoxin chemotypes. Above the CMC, the aggregate concentration increases and the monomer concentration remains constant. (**c**) Aggregate structures adopted by enterobacterial lipid A and LPS are complex hexagonal inverted (H_{II}) or cubic (Q) lipid phases. Lamellar bilayer or multilamellar (L) lipid phases are frequently observed for LPS with a reduced number of acyl chains. (**d**) Concept of the endotoxic conformation: Correlation of the aggregate structure and biological activity. Biological activity depends on the occurrence of hexagonal inverted or cubic lipid phases with a conical molecular geometry and a positive tilt angle of lipid A. LPS and lipid A with a cylindrical geometry and low or no backbone tilt angle form lamellar bilayer structures. These lipids do not activate the signaling receptor cascade but may be potent antagonists of TLR4 activation

small extrinsic changes such as hydration, ions, or temperature. For a determination of the aggregate structures, physical techniques such as small-angle scattering with X-rays (SAXS) or neutrons (SANS) must be applied.

The state of fluidity of the acyl chains directly affects the space coverage of the hydrophobic moiety and thus has a large impact on the occurrence of cubic and inverted aggregate structures that require conical molecules. Basically, two states of acyl chain fluidity can be adopted, the gel (β) and the liquid-crystalline (α) phase. In the gel phase the ordered acyl chains are in the all-trans configuration, while the liquid-crystalline (fluid) phase is much less ordered due to the introduction of increasing amounts of *gauche*-conformers. Between these phases, a (pseudo) first-order transition can be observed at a glycolipid-specific temperature T_m. The value of T_m is governed by various parameters such as the length and the degree of saturation of the hydrocarbon chains, the head group and further substitutions, hydration of the glycolipid, and solution properties such as pH, ionic strength, and the presence of divalent cations. Phase transition can take place while maintaining the aggregate structure, but phase transition can also induce conversion of the aggregate structure. Of note, not all aggregate structures can occur in both phases, in particular the H_{II} and Q structures are commonly not observed below T_m.

The structural polymorphism of endotoxins is described for lipid A, LPS Re, and other rough mutant LPS as well as wild-type LPS from *Salmonella minnesota* and *Escherichia coli* presenting complete phase diagrams by varying the concentration of water, Mg^{2+} as important physiological cations, and temperature [26–28]. All enterobacterial LPS exhibit a gel to liquid crystalline phase transition at T_m = 30 to 36 °C, depending on the length of the carbohydrate chain, with the lowest values for LPS Re. Of note, enterobacterial lipid A has the highest T_m with values around 45 °C. Most importantly, the aggregate structure of LPS adopts mainly non-lamellar organizations, which can be assigned to aggregates with inverted hexagonal H_{II} or cubic symmetry, in particular for endotoxins with short carbohydrate chains (lipid A, LPS Re). From these data, a conformational concept of endotoxins was deduced: the lipid A part of LPS adopts a conical shape with a cross-section of the hydrocarbon chains being higher than that of the hydrophilic part. This concept is valid for LPS with hexaacylated lipid A. In contrast, in penta- und tetraacylated lipid A the cross-sections of both molecular parts are nearly identical, forming multilamellar aggregates (Fig. 1c).

3 Role of Aggregates in Biological Activity

A question that has been discussed quite controversially in the literature is the question of the biologically active unit of LPS, whether this is the aggregate or the monomer. Several lines of evidence support that aggregates are the physical entities that are targeted by the immunological LPS-binding proteins in serum and that are required in the first place for the activation of TLR4 downstream of the transport chain. An experimental approach that strongly supports that aggregates are a prerequisite for biological recognition is the separation of aggregates and monomers in a diffusion chamber. Challenge of mononuclear cells with monomeric or aggregated LPS demonstrated that monomers were not able to induce cell activation, whereas aggregated LPS at the same concentrations showed robust cytokine induction. This result was observed in the absence as well as in the presence of human serum or in the presence of LPS-binding protein [29]. Other studies showed that a lower state of LPS aggregation is associated with largely reduced mortality in a model of galactosamine-sensitized mice [30]. From these data, it can be concluded that LPS in the aggregated state is required for biological recognition.

Analysis of the aggregate structures of a large number of lipid A and LPS, including preparations isolated from bacteria and analogs generated by chemical synthesis, by small-angle X-Ray diffraction (SAXS) have revealed a striking correlation of aggregate structure and biological activity. The three-dimensional organization of lipid aggregates is tightly connected with their ability to activate or antagonize cell activation. Thus, lipid structures with a conical molecular shape that assemble into complex H_{II} or Q lipid phases are correlated with high biological activity. In contrast, lipid A structures with a cylindrical molecular shape form lamellar or multilamellar lipid phases that do not express biological activity, however, several of these compounds express antagonistic activity, that is, they are able to inhibit cell activation by endotoxins [31]. This finding is supported by data for a variety of lipid A samples from different enterobacterial strains, lipid A in different salt forms, monophosphoryl lipid A (MPLA), lipid A from non-enterobacterial sources, and synthetic lipid A variants [32–34].

An important aspect of LPS aggregation states is the accessibility of the phosphate groups that have been shown to be of particular importance for the expression of biological activity [32, 35]. The bis-phosphorylated diglucosamine backbone does not align perpendicular to the membrane normal but adopts a tilt angle, which can be determined by FTIR spectroscopy using attenuated total reflectance analysis with polarized light [36]. The tilt observed in the lipid A backbone leads to exposure of the 1-phosphate group to the water phase, whereas the 4′-phosphate group is tilted downward pointing to the hydrophobic core of the membrane. The different degrees of hydration of the two phosphates are reflected by the infrared absorption peaks of the anti-symmetric stretching vibrations of the PO_2^- groups that can be determined in the wavenumber range of 1300–1260 cm^{-1}. Comparing the biological activities of different lipid A and LPS aggregate preparations, a strong correlation of a positive tilt angle of ≥35° with the expression of biological activity of lipid A is observed, whereas lipid A aggregates with a low tilt angle of the backbone around 10–15° express low or no biological activity [23]. Thus, a conical molecular conformation with a tilted glucosamine backbone exposing one phosphate group is the optimal structure for the expression of biological activity. Interestingly, the phosphate groups of LPS can be replaced by carboxymethyl groups without changing its bioactivity, but a negative charge is mandatory [34].

According to these findings, the term "endotoxic conformation" was coined, which relates the aggregate structure, the molecular conformation of individual molecules within the aggregates, and the biological activity to activate cells via TLR4 [37] (Fig. 1d). Of note, this correlation could also be confirmed for another group of TLR ligands, the bacterial lipopeptides, which activate host cells through the TLR2 receptor. Although the di- or triacylated lipopeptides show a considerably lower degree of acylation than the lipid A portion of LPS, their biological activity could also be correlated to the molecular conformation within the aggregated state with conical molecules expressing high TLR2 activity and cylindrical molecules expressing antagonistic activity [38]. These findings underline the fundamental validity of the importance of aggregation structure in biological lipid recognition by TLR receptors. New developments in the biological and chemical synthesis of lipid A molecules will provide the opportunity to generate a wide variety of new structures [39, 40]. Such pipelines open the opportunity to feed a larger number of compounds into structure activity relationship (SAR) studies and may thus expedite the identification of inhibitors with optimized physical and biological behavior. This is especially important in view of the need for new lead structures for therapeutics to cope with inflammatory diseases.

Interaction of LPS aggregates with components in the blood circulation has been demonstrated to modulate the structure and also the harmful pathophysiology of endotoxin [41]. Thus, binding of aggregated LPS to lipoproteins represents an important pathway of detoxification. An interaction of LPS with lipoprotein particles in blood was already discovered long before TLR4 was identified as the LPS receptor. Studies on the LPS-binding protein (LBP) and soluble CD14 (sCD14) revealed that sequential interaction of LBP and soluble CD14 (sCD14) catalyze a transfer of LPS molecules from aggregates to high density lipoproteins (HDL) in serum [42, 43]. This lipid transport is achieved by the activity of LBP, by extraction of LPS molecules from aggregates and catalyzing the occurrence of an intermediate stage of LPS-sCD14 complexes. LBP shuttles LPS from LPS-sCD14 complexes to a variety of lipoprotein particles present in the circulation, such as HDL, low density lipoproteins (LDL), very low density lipoproteins (VLDL), and chylomicrons. These microparticles are targeted from the circulation to the liver with subsequent neutralization of the endotoxic activity [44]. Analysis of smooth (S)-LPS and rough (R)-LPS glycoforms demonstrated a rapid removal of R-LPS aggregates from the circulation, whereas S-LPS aggregates showed prolonged residence time in the serum in vivo [45]. Protective effects with increased survival have been reported for application of reconstituted HDL particles in murine models of polymicrobial sepsis induced by cecal ligation and puncture, in intraperitoneal sepsis induced by injection of *Escherichia coli*, as well as in a model of *Pseudomonas aeruginosa*–induced pneumonia. These in vivo studies demonstrate the potent anti-inflammatory and endotoxin detoxifying effects of HDL particles [46, 47].

Protein interaction with endotoxin aggregates has also been demonstrated to modulate biological activity by directly changing the 3D structure of LPS aggregates. In this context, biophysical analysis of the mode of action of antimicrobial peptides (AMP) has shown the capacity of AMPs to convert the physical organization of LPS to multi-lamellar aggregates, an effect that is directly correlated to the endotoxin-neutralizing capacity [48, 49]. Mechanistic insights were revealed in particular from studies on the LPS-neutralizing polypeptide Aspidasept and variants thereof [50–52]. A very interesting finding, first observed by Jack Levin and co-workers, is the intriguing capacity of hemoglobin (Hb) to increase endotoxin-induced biological activities [53]. The addition of cross-linked Hb to penta-acylated LPS and lipid A preparations with a very low biological activity led to a drastic increase in cytokine secretion, such as TNF-α in human mononuclear cells [54]. Structural analysis demonstrated that Hb converts LPS aggregates from a non-lamellar structure to cubic symmetry. This was accompanied by a considerable reduction of the size and number of the original aggregates. Similar effects were also observed for TLR2 activating lipopeptides, suggesting a general molecular mechanism of Hb on aggregated lipids [55, 56]. It must be emphasized that Hb does not change the chemical structure of LPS or lipid A. An interesting aspect of the Hb activity is the observation that Hb itself shows membrane activity toward host cell membrane models. Thus, besides the direct effects of Hb on the aggregation structure of LPS, effects of Hb on the organization of the host cell membrane and the assembly and activation of the TLR4 receptor complex might also be considered for further mechanistic studies.

Natural LPS preparations derived from bacteria express quite a complex composition. Biological LPS aggregates are heterogeneous mixtures of diverse chemical LPS structures, containing different chemotypes ranging from deep-rough (Re) over rough mutant (Rb-Ra) LPS up to S-LPS [57]. The lipid A structure of the different chemotypes present in S-LPS preparations from wild-type bacteria displays considerable heterogeneity with respect to the acylation patter. Thus, in S-LPS from *Salmonella abortus equi*, the rough fraction was found to contain the expected acylation pattern of ester- and amide-bound 3-OH-14:0, whereas in the smooth fraction, a significant part of ester- as well as amino-linked acyl chains was absent [58]. An important observation in this context is the finding that the immunological active fraction of S-LPS is the R-chemotype fraction of LPS [59, 60].

Another aspect of the frequently observed presence of under-acylated lipid A structures such as pentaacyl and tetraacyl lipid A in natural LPS preparations is the low impact of these lipid species on the overall biological activity of the preparation. Surprisingly, these molecular species do not appear to express their antagonistic potential even when being present in amounts of up to 20% in the aggregates of biologically active LPS. Instead, mixing experiments demonstrated that the admixture of 10–20 Mol% of the synthetic antagonist 406 in aggregates composed of the synthetic lipid A compound 506 rather enhanced the biological activity. For the antagonistic glycolipid cardiolipin, similar results were obtained with up to 50 mole% of cardiolipin [29]. When the antagonists were not present in the lipid A aggregates but applied separately before stimulation with lipid A, complete inhibition of cell activation was observed. The finding that antagonistic compounds enhance endotoxic activity when present in the same aggregate indicate that the presentation of molecules in the aggregated state plays a decisive role for the molecular interaction of binding proteins and receptors.

The presented data support that in biological systems, LPS in the aggregated state is the physico-chemically relevant molecular state that is targeted by the participating transport molecules of the sequential recognition chain, which is discussed in Sect. 4.

4 From Aggregates to TLR4 Receptor Interaction

Biological recognition of LPS by TLR4 in the context of infections requires the extraction of the molecule from the bacterial cell envelope or from endotoxin aggregates to enable receptor binding. Within the family of Toll-like receptors, TLR4 has evolved the most complex cascade of using accessory proteins to enable the sensitive recognition of its ligand endotoxin. The accessory proteins comprise the LPS-binding protein (LBP), which is expressed by hepatocytes as an acute-phase protein and is highly upregulated upon infections [61, 62] and the soluble form of CD14, which is expressed as a glycosylphosphatidylinositol-anchored surface antigen mCD14 on monocytes and macrophages, and in lower amounts also on dendritic cells and neutrophils. CD14 is shed from the cell surface into serum, secreted from intracellular pools, and also produced as an acute-phase protein in the liver [63]. LBP binding to intact bacteria enhances phagocytosis. LBP can extract LPS monomers from the bacterial membrane [64, 65] and from endotoxin aggregates in solution [66]. A major function of the combined action of LBP and CD14 is to enable a highly sensitive activation of mononuclear cells [67]. Using ^{14}C- or ^{3}H-lipooligosaccharide (LOS) and LPS a sequential transport chain of LPS molecules from LBP to CD14, and subsequently to MD-2, reducing the binding affinity of the TLR4/MD2 complex to picomolar concentrations, was described [68, 69]. Of note, while metabolic radioactive labelling of LPS enables the sensitive detection of monomeric LPS, the emitted high radiation doses may induce cellular stress responses and could thus enhance also the cellular responsiveness. Reconstruction of the cascade by high resolution microscopy recently demonstrated on a molecular level that LBP acts as an accelerator of the initial process by allowing multiple rounds of LPS transfer to CD14 [70], further contributing to the extremely sensitive LPS recognition by the immune system.

From a biophysical perspective, thermodynamic considerations are highly relevant for the aggregation, disaggregation, and transport of LPS. Important insights into the transfer path of LPS were provided by molecular dynamic simulation analyses. The individual steps of LPS interaction were analyzed in a set of computational models of the bacterial outer membrane, the LPS aggregate, and complexes of CD14/LPS, MD-2/LPS, and CD14/TLR4/MD-2/LPS. The data obtained from these in silico studies revealed that channeling of the ligand along the receptor proteins binding lipid A with increasing affinity generates a thermodynamic funnel. The resulting energy gradient culminates in a terminal transfer of LPS to spontaneously assembled CD14/TLR4/MD-2 receptor complexes on the model of the host cell membrane [71]. In the bacterial membrane model, lipid A was retained with about 310 kJ mol^{-1} affinity, providing a high energy barrier for extraction. Divalent counterions such as Ca^{2+} and Mg^{2+} bridging the phosphate groups on the bacterial membrane contribute to this high energy barrier by tightly linking the negatively charged headgroups. The authors propose that LBP interaction disrupts this counterion barrier and thereby reduces the kinetic barrier for LPS extraction.

The TLR4/MD-2 complex is the signaling receptor for LPS at the cytoplasmic membrane [72–74]. Crystallographic data of complexes of the extracellular domain of TLR4 and MD-2 are the basis for the model that the receptor activation is enabled by the binding of a monomeric LPS molecule into the hydrophobic binding pocket of the accessory receptor protein MD-2, leading to the formation of TLR4/MD-2 heterodimers. Receptor dimerization is stabilized by the exposure of the acyl chain at position 2 at the lipid A at the surface of the MD-2 binding pocket, which forms together with the amino acid residue Phe126 of MD-2 a hydrophobic interaction patch for the adjacent TLR4 ectodomain. The complex is further stabilized by ionic interaction of the lipid A phosphate group in 4′ position with a cluster of positively charged residues in TLR4 and MD-2 [75, 76]. The dimeric receptor state is the platform that activates intracellular signaling via the engagement of adaptor protein myeloid differentiation primary response protein 88 (MYD88) assembly to large signaling platforms, the "Myddosome" complex activating the NF-kB pathway. Association of the TIR domain-containing adaptor protein inducing IFNβ (TRIF) to the TIR domains assembles the "Triffosome" platforms activating the interferon-regulatory factor (IRF)7-pathway [5]. Single molecule data obtained from quantitative super-resolution microscopy studies provide refined data about the receptor assembly. Krüger et al. revealed that the cell surface receptor complex of TLR4/MD-2 is present to about 50% in a monomeric state TLR4/MD-2 and to 50% in a dimeric state in unstimulated cells, demonstrating an intrinsic propensity of TLR4/MD-2 to dimerize. This ratio was dramatically shifted upon stimulation with LPS, leading to a large increase of the fraction of dimeric receptor state to about 75% [77]. A limitation of such investigations that has to be considered is the overexpression of TLR4_mEos2 receptors that were transfected in hamster embryonic kidnek cell line HEK293 in order to enable their detection in super resolution microscopy. The observation of dimerization of a significant pool of receptors was associated with a substantial degree of NF-κB activation even in the absence of LPS stimulation, supporting that dimerization and activation of TLR4 are closely related processes. Of note, data from molecular dynamic simulations on the thermodynamics involved in receptor ligand interaction suggest that the final step of LPS binding occurs to the preassembled TLR4/MD-2 dimeric receptors. This is in contrast to the mechanistic model that LPS induces receptor dimerization. Instead, it would be in accordance with the assumption that LPS binding leads to a stabilization of spontaneously formed receptor dimers [71].

5 Non-LPS Ligands of TLR4: Aggregate States and Biological Activity

Targeting of TLR4 by non-endotoxin compounds is of particular interest for therapeutic immune modulation of the receptor. The strong pathophysiology that can be elicited by TLR4 activation and the lack of treatment options for patients has attracted particular attention for the development of inhibitors. A successful line of development is based on lead structures of endotoxins with a naturally occurring low endotoxicity. LPS from the non-pathogenic bacterium *Rhodobacter sphaeroides* (Rs) LPS was discovered as a potent inhibitor of TLR4 activation by LPS. Rs lipid A became a lead structure for synthetic TLR4 antagonists. The compound Eritoran tetrasodium (E5564), generated by EISAI Inc. (Andover, USA), showed high antagonistic activity in murine and human cells while demonstrating improved stability, pharmacology, and pharmacokinetics [78–80]. The crystal structure of Eritoran bound to the mouse TLR4/MD-2 complex revealed the geometry, hydrophobic volume inside the binding cavity of MD-2 and the location of electrostatic interactions of the phosphate groups with positively charged patches at the rim of the hydrophobic pocket and the adjacent TLR4 molecule [81]. The visualization of the molecular geometry and the localization of the molecular interfaces greatly increase our understanding of the mode of action of the antagonist. One very interesting approach by EISAI Inc. was the generation of non-LPS compounds that mimic the physico-chemical characteristics of lipid A structures. They synthesized phospholipids with six acyl chains and two phosphates linked by a serine-like backbone, the latter with a spacer allowing to vary the length and the volume of the molecular backbone. By this strategy, lipid A mimicry molecules with different critical packing shapes were generated. Structural analysis of the molecular geometry of the compounds in the aggregates state revealed a conical molecular geometry for the compound with the smallest backbone (EISAI 803022), which expressed high biological activity. In contrast, compounds with a long spacer at the backbone expressed cylindrical molecular shape and were found more or less devoid of biological activity. The applicability of this structure activity correlation for molecules with different chemistry is referred to as "the generalized endotoxic principle" [38, 82, 83]. These findings clearly demonstrate the potency to develop non-LPS compounds as modulators for TLR4 by physico-chemical mimicry.

New approaches of targeting TLR4 pursue a strategy of simplified molecules for drug development. Synthetic disaccharide-based anionic amphiphiles were reported by Borio et al. as inhibitors of LPS-induced inflammation. The compounds were designed on the basis of optimizing the orientation and torsion of the MD-2 interacting groups. They show potent antagonistic activity at micromolar concentrations in human and murine macrophages. Structure–activity relationship studies and molecular dynamic simulation of the interaction with MD-2 were used to select two compounds with optimized properties as lead structures for future studies [84]; however, information on the molecular geometry and aggregation behavior is not yet available. Another example of synthetic TLR4 modulators is based on monosaccharide scaffolds. IAXO102 is a cationic antagonist inhibiting TLR4 activation in human cells and in an in vivo model of murine sepsis [85]. The FP series of anionic monosaccharide-based synthetic compounds was developed by computational approach to optimize docking of the lipids to the hydrophobic pocket of MD-2. For compound FP7 an insertion of the acyl chains into the hydrophobic cavity of MD-2 is demonstrated [86]. Antagonistic activity on LPS-induced activation of cytokines and chemokines is shown in human monocytes and dendritic cells [87]. A set of FP7 variants with variations of length of the acyl chains was analyzed in detail for physico-chemical properties such as solubility, structure, and phase-behavior in the aggregated state. The occurrence of mixed lamellar/nonlamellar or pure lamellar aggregate structures was observed, consistent with their low biological activity. FP compounds bind to MD-2 at μmolar concentrations and the most potent antagonistic activity on LPS-mediated TLR4/MD-2 activation in HEK cells was reported for FP7 with an IC50 of 2.0 μM and FP12 with 0.63 μM [88]. In addition, FP7, like Eritoran, demonstrated benefit in non-LPS inflammation in a murine model of lethal influenza infection, supporting that targeting TLR4 is not restricted to LPS-driven bacterial pathologies [87, 89].

6 LPS Aggregates in the Activation of Non-TLR4 Receptors

TLR4-independent recognition of LPS is a recently discovered alternative pathway enabling an immunological response to cytosolic LPS. Fifteen years after the identification of TLR4 as the LPS receptor [72], a role of the human caspases-4/-5 and the murine caspase-11 in sensing LPS in the cytoplasm of host cells was reported. These caspases were demonstrated to induce non-canonical inflammasome activation in response to LPS, leading to the production of the cytokines IL-1β and IL-18 and the induction of a pro-inflammatory type of lysis-induced cell death named pyroptosis [90, 91]. Activation of the intracellular caspase pathway by LPS is accompanied by a strong pathophysiology. Studies in TLR4-deficient mice provided evidence for TLR4 independent sepsis driven via non-canonical inflammasome activation [91]. LPS recognition by caspases was shown to involve direct binding of LPS to caspase-4/5/11 in vitro [92]. The binding of LPS to the caspase is mediated by the N-terminal caspase-activation and recruitment domain (CARD) domain. LPS-binding to the CARD domain induces oligomerization and is required for catalytic activity and biological activity of the caspase complexes. The molecular mechanisms involved in caspase activation by LPS are however not yet fully understood. A central question to be resolved is the pathway of cytosolic LPS delivery. Preparations of purified LPS that are highly potent in activating TLR4/MD-2 do not initiate the inflammasome pathway; however, specific delivery of the LPS into the cytosol is required. This delivery can be achieved experimentally by electroporation of cells or by transfection of cells using lipophilic cationic chelating reagents to complex LPS, both techniques representing non-physiologic artificial pathways to induce caspase-dependent inflammasome and IL-1β/IL-18 activation. Alternatively, in tissue culture settings the delivery of LPS by co-administration with cholera toxin B (CTB) mediates uptake via ganglioside M-1 (GM-1) and enables inflammasome activation [91, 92]. The discovery that LPS as part of outer membrane vesicles (OMVs) released from Gram-negative bacteria activate the non-canonical inflammasome pathway provided first evidence for a physiologic system of LPS transport to the intracellular compartment. The release of OMVs is a process observed in basically all Gram-negative bacteria and represents a vital sign of living bacteria [93, 94]. The uptake of OMVs involves phagocytosis and endosomal uptake [95, 96]. The mechanism of transfer or release of LPS from the vesicles into the cytosol is, however, unclear. Biophysical studies on OMVs demonstrated that in contrast to LPS, the LPS-containing bacterial vesicles can fuse with host cells, suggesting an intrinsic property for entry into the host cell cytoplasm [97]. Recent reports also indicate specific pathways for transport of LPS into the cytosol, candidates including the LPS-binding protein [98] and the high mobility group box 1 (HMGB-1) protein that mediates LPS transport via the RAGE receptor [99]. Involvement of a TLR4 – TIR-domain containing adapter-inducing interferon-β (TRIF) – guanylate-binding protein 2 (GBP2) pathway has been shown to enable the release of OMV-delivered LPS into the cytosol-inducing caspase activation in a murine model [100–102]. Since caspase regulation and activation displays a high diversity between different cell types and between murine and human cell systems, further studies will be needed to specify the role of different pathways. Considering the complex regulation of the TLR4 activation cascade, it is highly likely that not one but several pathways for LPS delivery may be in action to ensure the sensitive recognition of LPS.

The biophysical entity of caspase activation is also under debate. Biochemical studies demonstrated high affinity binding of purified caspase-4 to highly aggregated LPS and LPS-containing bacterial membrane vesicles. The resulting complexes did not indicate 1:1 molecular complexes but rather suggested an assembly of the caspase-4 at the LPS-containing membrane surfaces as a mechanism of activation [103]. Oligomerization of caspase by LPS as a critical step for caspase activation was also demonstrated in an independent study [92]. In contrast, another study showed by analytical ultracentrifugation and electron microscopy that binding of caspase-4 to large LPS aggregates induces disaggregation of LPS to low molecular weight caspase-4/LPS complexes [104]. Interestingly, for the caspase activation by particular LPS structures, also species-specific differences between human and murine caspases were reported [105]. Also in this relatively new field of LPS recognition by cytoplasmic caspases, it is apparent that the interaction of LPS with host cell membranes, transport proteins, and intracellular receptors is highly governed by biophysical mechanisms. Knowledge on the role of the aggregation state, the molecular geometry, and presentation of chemical groups of LPS to this group of receptors will be important to fully reveal their mode of activation.

7 Conclusions

The physico-chemistry of endotoxin is highly important for a mechanistic understanding of TLR4 activation by its ligand and also the basis for the development of new classes of immune-modulating compounds. A complete understanding of the molecular process of TLR4 activation has to consider the complex supramolecular structure, the variety of phase states and phase transitions occurring under different conditions, as well as the surface forces and thermodynamics governing the process of LPS recognition by TLR4. Future approaches for the discovery of new TLR4-interacting compounds should therefore consider not only the optimal receptor docking but also the physico-chemistry involved in LPS recognition.

References

1. Rietschel ET, Brade L, Schade U, et al. Bacterial lipopolysaccharides: relationship of structure and conformation to endotoxic activity, serological specificity and biological function. Adv Exp Med Biol. 1990;256:81–99.
2. Beutler B, Rietschel ET. Innate immune sensing and its roots: the story of endotoxin. Nat Rev Immunol. 2003;3:169–76.
3. Sharif O, Bolshakov VN, Raines S, Newham P, Perkins ND. Transcriptional profiling of the LPS induced NF-kappaB response in macrophages. BMC Immunol. 2007;8:1.
4. Fitzgerald KA, Rowe DC, Barnes BJ, Caffrey DR, Visintin A, Latz E, Monks B, Pitha PM, Golenbock DT. LPS-TLR4 signaling to IRF-3/7 and NF-kappaB involves the toll adapters TRAM and TRIF. J Exp Med. 2003;198:1043–55.
5. Gay NJ, Symmons MF, Gangloff M, Bryant CE. Assembly and localization of Toll-like receptor signalling complexes. Nat Rev Immunol. 2014;14:546–58.
6. Cohen J. The immunopathogenesis of sepsis. Nature. 2002;420:885–91.
7. van der Poll T, van de Veerdonk FL, Scicluna BP, Netea MG. The immunopathology of sepsis and potential therapeutic targets. Nat Rev Immunol. 2017;17:407–20.
8. Reyes M, Filbin MR, Bhattacharyya RP, et al. An immune-cell signature of bacterial sepsis. Nat Med. 2020;26:333–40.
9. Eng RH, Smith SM, Fan-Havard P, Ogbara T. Effect of antibiotics on endotoxin release from gram-negative bacteria. Diagn Microbiol Infect Dis. 1993;16:185–9.
10. Evans ME, Pollack M. Effect of antibiotic class and concentration on the release of lipopolysaccharide from Escherichia coli. J Infect Dis. 1993;167:1336–43.
11. Bucklin SE, Fujihara Y, Leeson MC, Morrison DC. Differential antibiotic-induced release of endotoxin from gram-negative bacteria. Eur J Clin Microbiol Infect Dis. 1994;13:S43–51.
12. Trautmann M, Heinemann M, Moricke A, Seidelmann M, Lorenz I, Berger D, Steinbach G, Schneider M. Endotoxin release due to ciprofloxacin measured by three different methods. J Chemother. 1999;11:248–54.
13. Parker SJ, Watkins PE. Experimental models of gram-negative sepsis. Br J Surg. 2001;88:22–30.
14. Galanos C, Freudenberg MA, Reutter W. Galactosamine-induced sensitization to the lethal effects of endotoxin. Proc Natl Acad Sci U S A. 1979;76:5939–43.
15. van Lier D, Geven C, Leijte GP, Pickkers P. Experimental human endotoxemia as a model of systemic inflammation. Biochimie. 2019;159:99–106.
16. Garidel P, Kaconis Y, Heinbockel L, Wulf M, Gerber S, Munk A, Vill V, Brandenburg K. Self-organisation, thermotropic and Lyotropic properties of glycolipids related to their biological implications. Open Biochem J. 2015;9:49–72.
17. Israelachvili JN. Intermolecular and surface forces. Burlington: Academic; 2011.
18. Maurer N, Glatter O, Hofer M. Determination of size and structure of lipid IVA vesicles by quasi-elastic light scattering and small-angle X-ray scattering. J Appl Crystallogr. 1991;24:832–5.
19. Aurell CA, Hawley ME, Wistrom AO. Direct visualization of gram-negative bacterial lipopolysaccharide self-assembly. Mol Cell Biol Res Commun. 1999;2:42–6.
20. Santos NC, Silva AC, Castanho MA, Martins-Silva J, Saldanha C. Evaluation of lipopolysaccharide aggregation by light scattering spectroscopy. Chembiochem. 2003;4:96–100.

21. Yu L, Tan M, Ho B, Ding JL, Wohland T. Determination of critical micelle concentrations and aggregation numbers by fluorescence correlation spectroscopy: aggregation of a lipopolysaccharide. Anal Chim Acta. 2006;556:216–25.
22. Sasaki H, White SH. Aggregation behavior of an ultra-pure lipopolysaccharide that stimulates TLR-4 receptors. Biophys J. 2008;95:986–93.
23. Brandenburg K, Andrä J, Müller M, Koch MH, Garidel P. Physicochemical properties of bacterial glycopolymers in relation to bioactivity. Carbohydr Res. 2003;338:2477–89.
24. Luzzati V, Vargas R, Mariani P, Gulik A, Delacroix H. Cubic phases of lipid-containing systems. Elements of a theory and biological connotations. J Mol Biol. 1993;229:540–51.
25. Luzzati V, Delacroix H, Gulik A, Gulik-Kryzwicki T, Mariani P, Vargas R. The cubic phase of lipids. In: Epand RM, editor. Current topics in membranes. San Diego: Academic; 1997.
26. Brandenburg K, Koch MH, Seydel U. Phase diagram of lipid A from Salmonella minnesota and Escherichia coli rough mutant lipopolysaccharide. J Struct Biol. 1990;105:11–21.
27. Brandenburg K, Koch MH, Seydel U. Phase diagram of deep rough mutant lipopolysaccharide from Salmonella minnesota R595. J Struct Biol. 1992;108:93–106.
28. Brandenburg K, Kusumoto S, Seydel U. Conformational studies of synthetic lipid A analogues and partial structures by infrared spectroscopy. Biochim Biophys Acta. 1997;1329:183–201.
29. Mueller M, Lindner B, Kusumoto S, Fukase K, Schromm AB, Seydel U. Aggregates are the biologically active units of endotoxin. J Biol Chem. 2004;279:26307–13.
30. Shnyra A, Hultenby K, Lindberg AA. Role of the physical state of Salmonella lipopolysaccharide in expression of biological and endotoxic properties. Infect Immun. 1993;61:5351–60.
31. Schromm AB, Brandenburg K, Loppnow H, Moran AP, Koch MH, Rietschel ET, Seydel U. Biological activities of lipopolysaccharides are determined by the shape of their lipid A portion. Eur J Biochem. 2000;267:2008–13.
32. Brandenburg K, Mayer H, Koch MH, Weckesser J, Rietschel ET, Seydel U. Influence of the supramolecular structure of free lipid A on its biological activity. Eur J Biochem. 1993;218:555–63.
33. Seydel U, Schromm AB, Blunck R, Brandenburg K. Chemical structure, molecular conformation, and bioactivity of endotoxins. Chem Immunol. 2000;74:5–24.
34. Seydel U, Schromm AB, Brade L, et al. Physicochemical characterization of carboxymethyl lipid A derivatives in relation to biological activity. FEBS J. 2005;272:327–40.
35. Bentala H, Verweij WR, Huizinga-Van der Vlag A, van Loenen-Weemaes AM, Meijer DK, Poelstra K. Removal of phosphate from lipid A as a strategy to detoxify lipopolysaccharide. Shock. 2002;18:561–6.
36. Seydel U, Oikawa M, Fukase K, Kusumoto S, Brandenburg K. Intrinsic conformation of lipid A is responsible for agonistic and antagonistic activity. Eur J Biochem. 2000;267:3032–9.
37. Seydel U, Brandenburg K, Rietschel ET. A case for an endotoxic conformation. Prog Clin Biol Res. 1994;388:17–30.
38. Schromm AB, Howe J, Ulmer AJ, et al. Physicochemical and biological analysis of synthetic bacterial lipopeptides: validity of the concept of endotoxic conformation. J Biol Chem. 2007;282:11030–7.
39. Scott AJ, Oyler BL, Goodlett DR, Ernst RK. Lipid A structural modifications in extreme conditions and identification of unique modifying enzymes to define the Toll-like receptor 4 structure-activity relationship. Biochim Biophys Acta Mol Cell Biol Lipids. 2017;1862:1439–50.
40. Sankaranarayanan K, Antaris XX, Palanski BA, El Gamal A, Kao CM, Fitch WL, Fischer CR, Khosla C. Tunable enzymatic synthesis of the Immunomodulator lipid IVA to enable structure-activity analysis. J Am Chem Soc. 2019;141:9474–8.
41. Andrä J, Gutsmann T, Mueller M, Schromm AB. Interactions between lipid A and serum proteins. Adv Exp Med Biol. 2010;667:39–51.
42. Wurfel MM, Hailman E, Wright SD. Soluble CD14 acts as a shuttle in the neutralization of lipopolysaccharide (LPS) by LPS-binding protein and reconstituted high density lipoprotein. J Exp Med. 1995;181:1743–54.

43. Yu B, Hailman E, Wright SD. Lipopolysaccharide binding protein and soluble CD14 catalyze exchange of phospholipids. J Clin Invest. 1997;99:315–24.
44. Vreugdenhil AC, Rousseau CH, Hartung T, Greve JW, van't Veer C, Buurman WA. Lipopolysaccharide (LPS)-binding protein mediates LPS detoxification by chylomicrons. J Immunol. 2003;170:1399–405.
45. Sali W, Patoli D, Pais de Barros JP, et al. Polysaccharide chain length of lipopolysaccharides from Salmonella Minnesota is a determinant of aggregate stability, plasma residence time and proinflammatory propensity in vivo. Front Microbiol. 2019;10:1774.
46. Guo L, Ai J, Zheng Z, Howatt DA, Daugherty A, Huang B, Li XA. High density lipoprotein protects against polymicrobe-induced sepsis in mice. J Biol Chem. 2013;288:17947–53.
47. Tanaka S, Geneve C, Zappella N, et al. Reconstituted high-density lipoprotein therapy improves survival in mouse models of Sepsis. Anesthesiology. 2020;132:825–38.
48. Andrä J, Howe J, Garidel P, et al. Mechanism of interaction of optimized Limulus-derived cyclic peptides with endotoxins: thermodynamic, biophysical and microbiological analysis. Biochem J. 2007;406:297–307.
49. Brandenburg K, Andrä J, Garidel P, Gutsmann T. Peptide-based treatment of sepsis. Appl Microbiol Biotechnol. 2011;90:799–808.
50. Kaconis Y, Kowalski I, Howe J, et al. Biophysical mechanisms of endotoxin neutralization by cationic amphiphilic peptides. Biophys J. 2011;100:2652–61.
51. Heinbockel L, Sanchez-Gomez S, Martinez de Tejada G, et al. Preclinical investigations reveal the broad-spectrum neutralizing activity of peptide Pep19-2.5 on bacterial pathogenicity factors. Antimicrob Agents Chemother. 2013;57:1480–7.
52. Correa W, Heinbockel L, Behrends J, et al. Antibacterial action of synthetic antilipopolysaccharide peptides (SALP) involves neutralization of both membrane-bound and free toxins. FEBS J. 2019;286:1576–93.
53. Su D, Roth RI, Levin J. Hemoglobin infusion augments the tumor necrosis factor response to bacterial endotoxin (lipopolysaccharide) in mice. Crit Care Med. 1999;27:771–8.
54. Brandenburg K, Garidel P, Andrä J, Jürgens G, Mueller M, Blume A, Koch MH, Levin J. Cross-linked hemoglobin converts endotoxically inactive pentaacyl endotoxins into a physiologically active conformation. J Biol Chem. 2003;278:47660–9.
55. Jürgens G, Mueller M, Koch MH, Brandenburg K. Interaction of hemoglobin with enterobacterial lipopolysaccharide and lipid A. Physicochemical characterization and biological activity. Eur J Biochem. 2001;268:4233–42.
56. Howe J, Richter W, Hawkins L, et al. Hemoglobin enhances the biological activity of synthetic and natural bacterial (endotoxic) virulence factors: a general principle. Med Chem. 2008;4:520–5.
57. Rietschel ET, Brade H, Holst O, et al. Bacterial endotoxin: chemical constitution, biological recognition, host response, and immunological detoxification. Curr Top Microbiol Immunol. 1996;216:39–81.
58. Jiao BH, Freudenberg M, Galanos C. Characterization of the lipid A component of genuine smooth-form lipopolysaccharide. Eur J Biochem. 1989;180:515–8.
59. Huber M, Kalis C, Keck S, et al. R-form LPS, the master key to the activation of TLR4/MD-2-positive cells. Eur J Immunol. 2006;36:701–11.
60. Pupo E, Lindner B, Brade H, Schromm AB. Intact rough- and smooth-form lipopolysaccharides from Escherichia coli separated by preparative gel electrophoresis exhibit differential biologic activity in human macrophages. FEBS J. 2013;280:1095–111.
61. Schumann RR, Leong SR, Flaggs GW, Gray PW, Wright SD, Mathison JC, Tobias PS, Ulevitch RJ. Structure and function of lipopolysaccharide binding protein. Science. 1990;249:1429–31.
62. Schumann RR, Kirschning CJ, Unbehaun A, Aberle HP, Knope HP, Lamping N, Ulevitch RJ, Herrmann F. The lipopolysaccharide-binding protein is a secretory class 1 acute-phase protein whose gene is transcriptionally activated by APRF/STAT/3 and other cytokine-inducible nuclear proteins. Mol Cell Biol. 1996;16:3490–503.

63. Bas S, Gauthier BR, Spenato U, Stingelin S, Gabay C. CD14 is an acute-phase protein. J Immunol. 2004;172:4470–9.
64. Vesy CJ, Kitchens RL, Wolfbauer G, Albers JJ, Munford RS. Lipopolysaccharide-binding protein and phospholipid transfer protein release lipopolysaccharides from gram-negative bacterial membranes. Infect Immun. 2000;68:2410–7.
65. Kitchens RL, Thompson PA. Modulatory effects of sCD14 and LBP on LPS-host cell interactions. J Endotoxin Res. 2005;11:225–9.
66. Hailman E, Lichenstein HS, Wurfel MM, Miller DS, Johnson DA, Kelley M, Busse LA, Zukowski MM, Wright SD. Lipopolysaccharide (LPS)-binding protein accelerates the binding of LPS to CD14. J Exp Med. 1994;179:269–77.
67. Wright SD, Ramos RA, Tobias PS, Ulevitch RJ, Mathison JC. CD14, a receptor for complexes of lipopolysaccharide (LPS) and LPS binding protein. Science. 1990;249:1431–3.
68. Gioannini TL, Teghanemt A, Zhang D, Coussens NP, Dockstader W, Ramaswamy S, Weiss JP. Isolation of an endotoxin-MD-2 complex that produces Toll-like receptor 4-dependent cell activation at picomolar concentrations. Proc Natl Acad Sci U S A. 2004;101:4186–91.
69. Prohinar P, Re F, Widstrom R, Zhang D, Teghanemt A, Weiss JP, Gioannini TL. Specific high affinity interactions of monomeric endotoxin protein complexes with Toll-like receptor 4 ectodomain. J Biol Chem. 2007;282:1010–7.
70. Ryu JK, Kim SJ, Rah SH, et al. Reconstruction of LPS transfer Cascade reveals structural determinants within LBP, CD14, and TLR4-MD2 for efficient LPS recognition and transfer. Immunity. 2017;46:38–50.
71. Huber RG, Berglund NA, Kargas V, Marzinek JK, Holdbrook DA, Khalid S, Piggot TJ, Schmidtchen A, Bond PJ. A thermodynamic funnel drives bacterial lipopolysaccharide transfer in the TLR4 pathway. Structure. 2018;26(1151–1161):e4.
72. Poltorak A, He X, Smirnova I, et al. Defective LPS signaling in C3H/HeJ and C57BL/10ScCr mice: mutations in Tlr4 gene. Science. 1998;282:2085–8.
73. Shimazu R, Akashi S, Ogata H, Nagai Y, Fukudome K, Miyake K, Kimoto M. MD-2, a molecule that confers lipopolysaccharide responsiveness on Toll-like receptor 4. J Exp Med. 1999;189:1777–82.
74. Schromm AB, Lien E, Henneke P, et al. Molecular genetic analysis of an endotoxin nonresponder mutant cell line: a point mutation in a conserved region of MD-2 abolishes endotoxin-induced signaling. J Exp Med. 2001;194:79–88.
75. Park BS, Song DH, Kim HM, Choi BS, Lee H, Lee JO. The structural basis of lipopolysaccharide recognition by the TLR4-MD-2 complex. Nature. 2009;458:1191–5.
76. Ohto U, Fukase K, Miyake K, Shimizu T. Structural basis of species-specific endotoxin sensing by innate immune receptor TLR4/MD-2. Proc Natl Acad Sci U S A. 2012;109:7421–6.
77. Krüger CL, Zeuner MT, Cottrell GS, Widera D, Heilemann M. Quantitative single-molecule imaging of TLR4 reveals ligand-specific receptor dimerization. Sci Signal. 2017;10:eaan1308.
78. Hawkins LD, Christ WJ, Rossignol DP. Inhibition of endotoxin response by synthetic TLR4 antagonists. Curr Top Med Chem. 2004;4:1147–71.
79. Mullarkey M, Rose JR, Bristol J, et al. Inhibition of endotoxin response by e5564, a novel Toll-like receptor 4-directed endotoxin antagonist. J Pharmacol Exp Ther. 2003;304:1093–102.
80. Rossignol DP, Wasan KM, Choo E, Yau E, Wong N, Rose J, Moran J, Lynn M. Safety, pharmacokinetics, pharmacodynamics, and plasma lipoprotein distribution of eritoran (E5564) during continuous intravenous infusion into healthy volunteers. Antimicrob Agents Chemother. 2004;48:3233–40.
81. Kim HM, Park BS, Kim JI, et al. Crystal structure of the TLR4-MD-2 complex with bound endotoxin antagonist Eritoran. Cell. 2007;130:906–17.
82. Seydel U, Hawkins L, Schromm AB, Heine H, Scheel O, Koch MH, Brandenburg K. The generalized endotoxic principle. Eur J Immunol. 2003;33:1586–92.
83. Brandenburg K, Hawkins L, Garidel P, Andra J, Muller M, Heine H, Koch MH, Seydel U. Structural polymorphism and endotoxic activity of synthetic phospholipid-like amphiphiles. Biochemistry. 2004;43:4039–46.

84. Borio A, Holgado A, Garate JA, Beyaert R, Heine H, Zamyatina A. Disaccharide-based anionic Amphiphiles as potent inhibitors of lipopolysaccharide-induced inflammation. ChemMedChem. 2018;13:2317–31.
85. Piazza M, Rossini C, Della Fiorentina S, Pozzi C, Comelli F, Bettoni I, Fusi P, Costa B, Peri F. Glycolipids and benzylammonium lipids as novel antisepsis agents: synthesis and biological characterization. J Med Chem. 2009;52:1209–13.
86. Cighetti R, Ciaramelli C, Sestito SE, et al. Modulation of CD14 and TLR4.MD-2 activities by a synthetic lipid A mimetic. Chembiochem. 2014;15:250–8.
87. Perrin-Cocon L, Aublin-Gex A, Sestito SE, et al. TLR4 antagonist FP7 inhibits LPS-induced cytokine production and glycolytic reprogramming in dendritic cells, and protects mice from lethal influenza infection. Sci Rep. 2017;7:40791.
88. Facchini FA, Zaffaroni L, Minotti A, et al. Structure-activity relationship in monosaccharide-based toll-like receptor 4 (TLR4) antagonists. J Med Chem. 2018;61:2895–909.
89. Shirey KA, Lai W, Scott AJ, et al. The TLR4 antagonist Eritoran protects mice from lethal influenza infection. Nature. 2013;497:498–502.
90. Kayagaki N, Wong MT, Stowe IB, et al. Noncanonical inflammasome activation by intracellular LPS independent of TLR4. Science. 2013;341:1246–9.
91. Hagar JA, Powell DA, Aachoui Y, Ernst RK, Miao EA. Cytoplasmic LPS activates caspase-11: implications in TLR4-independent endotoxic shock. Science. 2013;341:1250–3.
92. Shi J, Zhao Y, Wang Y, Gao W, Ding J, Li P, Hu L, Shao F. Inflammatory caspases are innate immune receptors for intracellular LPS. Nature. 2014;514:187–92.
93. Kulp A, Kuehn MJ. Biological functions and biogenesis of secreted bacterial outer membrane vesicles. Annu Rev Microbiol. 2010;64:163–84.
94. Kaparakis-Liaskos M, Ferrero RL. Immune modulation by bacterial outer membrane vesicles. Nat Rev Immunol. 2015;15:375–87.
95. Bomberger JM, Maceachran DP, Coutermarsh BA, Ye S, O'Toole GA, Stanton BA. Long-distance delivery of bacterial virulence factors by Pseudomonas aeruginosa outer membrane vesicles. PLoS Pathog. 2009;5:e1000382.
96. Vanaja SK, Russo AJ, Behl B, Banerjee I, Yankova M, Deshmukh SD, Rathinam VAK. Bacterial outer membrane vesicles mediate cytosolic localization of LPS and Caspase-11 activation. Cell. 2016;165:1106–19.
97. Jäger J, Keese S, Roessle M, Steinert M, Schromm AB. Fusion of Legionella pneumophila outer membrane vesicles with eukaryotic membrane systems is a mechanism to deliver pathogen factors to host cell membranes. Cell Microbiol. 2015;17:607–20.
98. Kopp F, Kupsch S, Schromm AB. Lipopolysaccharide-binding protein is bound and internalized by host cells and colocalizes with LPS in the cytoplasm: implications for a role of LBP in intracellular LPS-signaling. Biochim Biophys Acta. 2016;1863:660–72.
99. Deng M, Tang Y, Li W, et al. The endotoxin delivery protein HMGB1 mediates Caspase-11-dependent lethality in Sepsis. Immunity. 2018;49(740–753):e7.
100. Santos JC, Dick MS, Lagrange B, et al. LPS targets host guanylate-binding proteins to the bacterial outer membrane for non-canonical inflammasome activation. EMBO J. 2018;37:e98089.
101. Gu L, Meng R, Tang Y, et al. Toll-like receptor 4 Signaling licenses the cytosolic transport of lipopolysaccharide from bacterial outer membrane vesicles. Shock. 2019;51:256–65.
102. Rathinam VAK, Zhao Y, Shao F. Innate immunity to intracellular LPS. Nat Immunol. 2019;20:527–33.
103. Wacker MA, Teghanemt A, Weiss JP, Barker JH. High-affinity caspase-4 binding to LPS presented as high molecular mass aggregates or in outer membrane vesicles. Innate Immun. 2017;23:336–44.
104. An J, Kim SH, Hwang D, Lee KE, Kim MJ, Yang EG, Kim SY, Chung HS. Caspase-4 disaggregates lipopolysaccharide micelles via LPS-CARD interaction. Sci Rep. 2019;9:826.
105. Lagrange B, Benaoudia S, Wallet P, et al. Human caspase-4 detects tetra-acylated LPS and cytosolic Francisella and functions differently from murine caspase-11. Nat Commun. 2018;9:242.

CD14: Not Just Chaperone, But a Key-Player in Inflammation

Marco Di Gioia and Ivan Zanoni

Abstract The cluster of differentiation 14 (CD14) is a receptor that helps the detection of microbial products, acting as an accessory molecule for several Toll-like receptors (TLRs). Recently, this view was challenged by the discovery that CD14 not only regulates TLR-ligand interactions but also autonomously controls inflammatory responses. Moreover, CD14 is also implicated in the recognition of host-derived tissue stress signals, such as non-enzymatically modified lipids and protein aggregates, playing homeostatic roles as well as initiating detrimental responses associated to chronic inflammation. Here we will discuss the ability of CD14 to bind different exogenous and endogenous ligands and how CD14 coordinates inflammatory responses.

Keywords CD14 · Pattern recognition receptor (PRR) · Toll-like receptor (TLR) · Co-receptor · Inflammation

1 Introduction

The immune system of mammals evolved to maintain their complex tissue and systemic homeostasis. The immune system functions as a "caretaker" to react against exogenous and endogenous stressors by activating appropriate effector responses. Immune cells are strategically distributed throughout the organism to detect the presence of invading microorganisms (exogenous stressors) as well as host-derived

M. Di Gioia (✉)
Division of Immunology, Boston Children's Hospital, Harvard Medical School, Boston, MA, USA
e-mail: marco.digioia@childrens.harvard.edu

I. Zanoni (✉)
Division of Immunology, Boston Children's Hospital, Harvard Medical School, Boston, MA, USA

Division of Gastroenterology, Boston Children's Hospital, Harvard Medical School, Boston, MA, USA
e-mail: ivan.zanoni@childrens.harvard.edu

C. Rossetti, F. Peri (eds.), *The Role of Toll-Like Receptor 4 in Infectious and Non Infectious Inflammation*, Progress in Inflammation Research 87,
https://doi.org/10.1007/978-3-030-56319-6_4

threats, such as the abnormal production and/or accumulation of self-biomolecules (endogenous stressors). The stressors are generally recognized by specialized receptors called pattern recognition receptors (PRRs) that trigger a context-dependent functional response to restore healthy physiological conditions. CD14 is a master regulator of the responses mediated by several PRRs, such as toll-like receptors (TLRs) TLR4 and TLR2, by modulating the specificity, affinity and accessibility for their ligands or controlling their subcellular localization. In addition to this "co-receptor" or "chaperone" functions, CD14 also exerts autonomous functions, by regulating efferocytotic/endocytic processes or by activating dedicated signaling pathways. In this chapter we will describe the ability of CD14 to function as a regulator of the inflammatory process elicited in response to exogenous and/or endogenous stimuli.

2 CD14 Overview

2.1 CD14 Structure

CD14 is a glycoprotein that belongs to the leucine-rich repeat (LRR) family. CD14 is anchored to the outer leaflet of the plasma membrane by a glycosylphosphatidylinositol (GPI) moiety, and it localizes in microdomains called lipid rafts [1, 2]. Alternatively, CD14 can be released in the blood and extravascular fluids via three different mechanisms: (1) the proteolytic removal of the C-terminal sequence necessary for GPI attachment [3, 4]; (2) escape from posttranslational modification [4]; and (3) direct GPI anchor cleavage by phospholipase D [5, 6]. When newly synthetized, the human CD14 includes: (i) a N-terminal signal peptide necessary to address the protein to the secretory pathway, which is cleaved during protein maturation; (ii) an extracellular domain composed of 5 α-helices and 11 β-strands, forming a horse shoe-shaped structure (53 kDa for membrane-bound CD14 and 48 kDa for soluble form of CD14 originated by proteolytic mechanisms [1]); and (iii) a C-terminal propeptide, removed in the mature configuration [7, 8] (Fig. 1).

A key structural element in both human and murine CD14 is the presence of an N-terminal pocket [7, 9] that is indispensable for the binding of acylated ligands, including lipopolysaccharide (LPS), a major component of gram-negative bacteria outer membrane and the first identified, and most studied, CD14 ligand [12]. This hydrophobic cavity includes four functional regions, designated as R1 to R4, respectively, that are involved in LPS binding and necessary for the activity of CD14 [13, 14]. Interestingly, the relatively big size of the pocket and the presence of hydrophobic aminoacidic clusters located just outside of the pocket entrance [7, 9] may explain the broad ligand specificity of CD14, although further studies are required to confirm this hypothesis.

Another essential feature of the structure of CD14 is the presence of specific basic and acidic patches that are important in protein–protein and ligand interactions. These patches allow CD14 to recover LPS from LPS-binding protein (LBP)

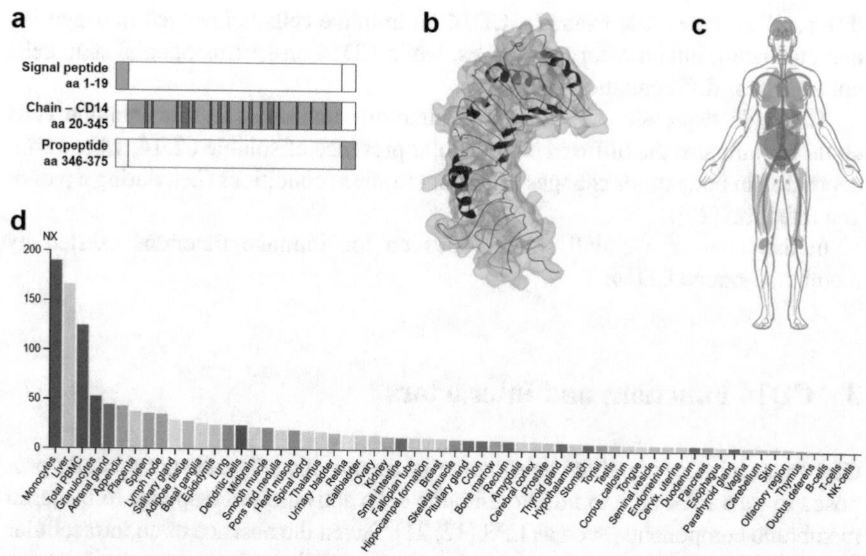

Fig. 1 Human CD14 structure and expression. (**a**) The CD14 protein aminoacidic scheme shows the processing events that lead to the formation of the mature form of human CD14. Secondary structure positions of α helices (red) and β strands (blue) are indicated [8, 9]. (**b**) 3D structure of human CD14 shows secondary structures (inner cartoon) and molecular surfaces [8, 9]. (**c**) Anatogram and (**d**) RNA expression overview of CD14 body distribution. Consensus normalized expression (NX) levels for 55 tissue types and 6 blood cell types were created by combining the data from three RNA-seq datasets (HPA, GTEx, and FANTOM5). Color-coding is based on tissue groups, each consisting of tissues with common functional features [10, 11]

[15, 16]. In particular, a C-terminal acidic sequence on CD14 concave surface interacts with conserved basic residues at the C-terminal tip of the LPS-LBP complex allowing the delivery of LPS from LBP to CD14. Then, acidic residues within the N-terminal portion of LBP and CD14 mediate the prompt dissociation of CD14 via an electrostatic repulsion [15].

2.2 *CD14 Expression*

Genome-wide analysis at the tissue and single cell level allowed to trace a map of the human transcriptome and proteome [10, 11]. These data confirmed previous observations [17], showing that, among immune cells, CD14 is selectively expressed by myeloid cells, such as monocytes and macrophages, granulocytes, and dendritic cells (DCs). CD14 is also present in non-immune cells, such as hepatocytes, and it is broadly distributed in other human tissues though at lower expression levels (Fig. 1). These data corroborate several reports that documented the presence of CD14 also in epithelia, smooth muscle, pancreatic islet cells, fibroblasts, adipocytes, and spermatozoa [18]. Membrane-bound CD14 plays different roles in different cell types. For example, CD14 on immune cells is involved in triggering and enhancing inflammatory responses, while CD14 on spermatogonial stem cells controls their differentiation [19].

The wide expression of CD14 by numerous immune and non-immune cells could explain also the diffused extracellular presence of soluble CD14, whose concentration in body fluids changes according to stress conditions (i.e., during a pathogen infection [20]).

In this chapter, we will solely focus on the immune functions exerted by membrane-bound CD14.

3 CD14 Functions and Interactors

CD14 was initially identified as a marker of monocytes [17] and subsequently proposed to be a PRR able to initiate immune cells activation in response to bacterial membrane components, such as LPS [12, 21]. Given the absence of an intracellular domain, CD14 was believed to work only in association with companion proteins endowed with signaling capacities. When the TLR4 was cloned and identified as the LPS receptor [22, 23], CD14 was believed to function solely as the TLR4 co-receptor. More recently, CD14 was found to interact with a plethora of ligands other than LPS, and that it not only serves as a co-receptor for several TLRs (TLR1/2/3/4/6), but it is also endowed with signaling capacities that impact the final outcome of the inflammatory response.

3.1 CD14 in LPS Signaling: TLR4-Dependent and TLR4-Independent Responses

CD14 plays a critical role in controlling the inflammatory responses triggered by gram-negative bacteria. Indeed, CD14-deficient mice are resistant to the endotoxic shock induced by LPS, *Escherichia coli* [24] or *Burkholderia pseudomallei* [25]. Also, functional polymorphisms within the human CD14 gene can differentially impact the induction of pro-inflammatory responses and the clinical prognosis of septic patients [26–32]. The roles played by CD14 during gram-negative bacterial infections are due to its capacity to regulate, both directly and indirectly, numerous aspects of the activity of the TLR4-Myeloid differentiation protein-2 (MD2) complex. Indeed, CD14 can: (1) dictate the access of the extracellular LPS to TLR4-MD2; (2) regulate the sub-cellular localization of LPS and TLR4-MD2; and (3) trigger a TLR4-independent, calcium-dependent response that selectively influences the production of inflammatory mediators and the cell survival.

3.1.1 CD14 Controls Extracellular LPS Capture, Sensing, and Signaling

LPS is released from the bacterial membrane in the form of micelles, or it can be actively secreted via the formation of outer membrane vesicles (OMVs) [33]. The latter can transport biochemical signals to other bacteria or host cells and they can directly deliver LPS in the cytosol of immune cells, where inflammatory caspases (caspase-11/4/5) serve as a specialized LPS receptor to induce the activation of the inflammasome and the production of bioactive interleukin-1β (IL-1β) and IL-18 [34]. In contrast to OMVs, the LPS contained in micelles requires the presence of accessory soluble proteins, such as LBP or High Mobility Group Box 1 (HMGB1), to be recognized by TLR4-MD2 or caspase-11/4/5 [15, 16, 35, 36]. While the transport of micelles to inflammatory caspases requires an endocytic process mediated by HMGB1 and by the receptor for advanced glycation end products (RAGE) [35], CD14 enhances the sensitivity of TLR4-MD2 up to picomolar concentrations of LPS [37] and regulates the subcellular location of the TLR4-MD2 complex (Fig. 2).

The transfer of LPS monomers from micelles to TLR4-MD2 via CD14 requires either LBP [15, 16] or HMGB1 [36]. The interactions between LBP or HMGB1 and CD14 form a "capture and concentration module" upstream of TLR4-MD2 that regulates the ligand availability. HMG1 and LBP transport LPS to CD14 from LPS micelles. Soluble LPS binding proteins start the process by binding LPS micelles and therefore catalyzing several rounds of LPS monomers transfer to CD14. Subsequently, the single LPS molecule bound to CD14 is transferred to MD2 with the assistance of LRR13-LRR15 domains of TLR4 that trigger the multimerization of TLR4-MD2 and its activation [15, 16]. Contextually to LPS presentation, CD14 also facilitates the relocation of TLR4-MD2 in lipid rafts, where multiple signaling molecules are recruited to contribute to cell activation [38]. These specialized plasma membrane regions contain dock sites for several interactors involved both in TLR4-depedent pathways (such as toll-interleukin 1 receptor domain containing adaptor protein [TIRAP] and myeloid differentiation primary response 88 [MyD88] adaptors, which leads to NF-kB activation and cytokine production [39]) as well as TLR4-independet effectors, such as specialized proteins for the subsequent internalization of the complex formed by LPS, CD14, and TLR4-MD2 (Fig. 2).

3.1.2 CD14 Controls LPS Endosomal Signaling

Once engaged by CD14, TLR4-MD2 undergoes an internalization process and moves in the endosomal compartment, where it triggers the TRIF-related adaptor molecule (TRAM) and TIR-domain-containing adapter-inducing interferon-β (TRIF)-dependent pathway, which sustains the activation of NF-κB and also induces the production of type I interferons (IFNs) [40–42].

The internalization of the complex containing LPS and TLR4-MD2 starts with the interaction between CD14 and MD2, which functions as a cargo-selection agent for TLR4 [14]. Notably, CD14 has an autonomous endocytic activity. In particular,

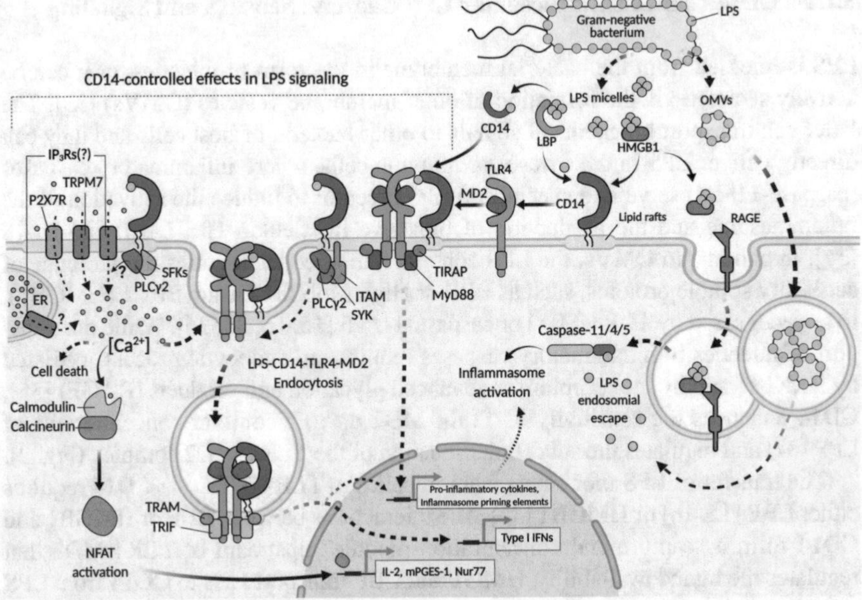

Fig. 2 LPS signaling. Extracellular gram-negative bacteria release LPS in the form of micelles or OMVs. OMVs and micelles containing LPS, the latter with the help of HMGB1-RAGE, can be delivered intracellularly, where LPS activates caspase-11/4/5-depedent responses (right). Soluble LPS-binding proteins allow CD14 to capture LPS monomers. CD14 increases the sensitivity of TLR4-MD2 for LPS and favors the re-location of the complex formed by LPS, CD14, TLR4-MD2 in the plasma membrane lipids rafts. Once in the lipid rafts, TLR4-MD2 starts TIRAP-MyD88-dependent responses. By activating ITAM-containing adaptors, SYK, and PLCγ2, CD14 also induces the endocytosis of LPS and TLR4-MD2. From endosomes TLR4-MD2 triggers the TRAM-TRIF pathway and thereby sustains the activation of NF-κB and the production of type I IFNs. Finally, CD14 activates calcium-dependent responses, inducing calcium fluxes from the extracellular space through PX27R, TRPM7, and possibly IP_3Rs. The increase in the local intracellular calcium concentration contributes to the internalization of LPS, CD14, and TLR4-MD2, induces phagocytes' cell death, and, in DCs, activates the NFAT-dependent signaling cascade

CD14 can activate two downstream systems that contribute to the endocytosis of TLR4-MD2: (1) CD14 activates the spleen tyrosine kinase (SYK) and phospholipase Cγ2 (PLCγ2) via immunoreceptor tyrosine-based activation motif (ITAM)-containing adaptors, such as DNAX-activating protein 12 (DAP12) and the γ-subunit of high-affinity IgE receptor (FcεRIγ). PLCγ2 breaks down the phosphatidylinositol 4,5-bisphosphate (PIP_2) present in the lipid rafts, thereby releasing TIRAP-MyD88 from TLR4 and triggering the formation of an early endosome [40, 42, 43]. (2) CD14 induces the opening of calcium channels, such as inositol-1,4,5-triphosphate receptors (IP_3Rs) [44] and transient receptor potential melastatin-like 7 (TRPM7) [45], which increase local calcium concentrations and assist the completion of the early endosome. Once in the endosome, the intracellular domain of TLR4 contacts the TRAM-TRIF complex and induces the second phase of the signaling, ultimately leading to the induction of type I IFNs production (Fig. 2). Notably type

I IFNs, besides playing anti-microbial roles during bacterial infections [46], also act on immune cells to regulate their metabolic program. In particular, type I IFNs control the production of succinate, itaconate, nitric oxide, and IL-10 [47–49]. Given that CD14 controls the endocytic process of TLR4-MD2 that leads to the production of type I IFNs, it will be important in the future to assess how CD14 directly or indirectly modulates the metabolic changes that follow LPS administration.

In agreement with the fundamental importance of CD14 in regulating the immune response against gram-negative bacteria, host-adapted pathogenic bacteria and commensals have developed several strategies to selectively avoid CD14 engagement or to uncouple CD14 endocytosis from TLR4 internalization. For example, *Bacteroides thetaiotaomicron*, an opportunist pathogen found in the human gut, presents a hypophosphorylated LPS form that evades CD14-dependent events, preventing inflammation and increasing its persistence in the host [14].

3.1.3 CD14 Controls LPS-Induced Calcium-Dependent Pathways in a TLR4-Independent Manner

In phagocytes, CD14 also acts as an autonomous LPS-receptor able to induce calcium-dependent responses that control the activation of specific transcription factors and/or cell viability [45, 50, 51].

In DCs, after LPS stimulation, CD14 triggers Src family kinases (SFKs) and PLCγ2 activation, resulting an extracellular Ca^{2+} influx [50]. The exact nature of the channel responsible for this process is still unknown, although the production of IP_3 mediated by PLCγ2 suggested the possible involvement of IP_3Rs [44, 50]; furthermore CD14 has been shown to interact with purinergic type 2 receptor family X7 receptor (P2X7R) [51] and TRPM7 [45] in alveolar macrophages (AMs) and bone marrow–derived macrophages (BMDMs), respectively. The possibility that CD14 acts simultaneously on multiple calcium channels and/or works in a cell-specific manner remains to be tested. The increase in the cytosolic calcium level promotes the activation of calcineurin and, consequently, the nuclear translocation of members of the nuclear factor of activated T-cell (NFAT) transcription factor family (Fig. 2). The activation of NFAT, in association with the activity of transcription factors controlled by the activation of TLR4, controls the expression of specific genes such as IL-2, nuclear receptor subfamily 4 group A member 1 (NR4A1; also known as Nur77) [50], and microsomal prostaglandin (PG) E synthase 1 (mPGES-1) [52]. The regulation of these genes affects several DC-related functions. In particular, it has been shown that DC-derived IL-2 is required for natural killer (NK) cell activation [53], T-cell expansion and development of antigen-specific effector cells [54], T helper (Th) 17 cell functions [55], and regulatory T-cell (T_{reg}) homeostasis [56, 57]. IL-2 produced by $CD11c^{high}MHCII^{+}$ myeloid cells appears to be particularly relevant for the maintenance of intestinal immune homeostasis, by supporting T_{reg} induction and by suppressing excessive expansion of Th1/Th17 cells [58]. By controlling the expression of NR4A1, a pro-apoptotic factor [59], the CD14-NFAT pathway modulates the life cycle of DCs [60]. This was shown to be particularly relevant to restrict the duration of antigen presentation by DCs and to avoid autoimmune disease development [60]. Finally, the expression of mPGES-1 induced by the CD14-NFAT pathway is necessary for the biosynthesis of PGE_2. The DC-derived PGE_2 is implicated in skin edema formation and in the delivery of free antigens to draining lymph nodes, thereby regulating the antigen-specific adaptive immune responses [52].

In AMs, CD14 can favor the activation of the aforementioned P2X7R calcium channel by LPS. P2X7R contains a cytosolic C-terminal lipid interaction site [61] that CD14 makes accessible to LPS through the internalization of the ligand and the physical interaction between LPS-CD14 and P2X7R. Upon LPS recognition via CD14, P2X7R induces Ca^{2+} influx and ATP depletion, which leads to necrosis and the release of pro-IL-1α. IL-1α in turn stimulates endothelial cell activation allowing neutrophil recruitment and sustains LPS-induced acute lung injury (ALI) [51].

3.2 CD14-Dependent Recognition of Non-LPS Ligands and Their Interactors

As discussed in 1.1, CD14 has a large hydrophobic pocket able to accommodate lipid chains, which, besides LPS, allows CD14 to bind several exogenous and endogenous fatty acid moieties (Table 1). CD14 also interacts with non-lipidic biomolecules such as peptidoglycan and extracellular proteins that undergo co- and post-translational glycosylation processes, such as amyloid precursor protein and extracellular matrix components (Table 1). Indeed, it has been proposed that CD14 might function also as a lectin-like protein [123]. Ultimately, CD14 can interplay with ligands/companion proteins via electrostatic forces [15] (Table 1). A yeast two-hybrid screening on human liver proteome led to the identification of putative additional CD14 interactors (such as ARFGAP3, GJB2, IRAK3, POLR2J, PPP3R2, and TNFRSF1B, identified by [124]), while affinity capture-mass spectrometry identified AP3S1, OASL, PHF20L1, ARID2, SVIL, BUB1B, ICAM5, TAF1, ANXA7, PHF21A, PPP1R12B, HPS5, BCAS3, and EXO1 as other possible CD14 partners [125]. The broad binding capacities and the diffused distribution throughout the body of CD14 make its importance as a regulator of multiple signal cascades evident.

3.2.1 CD14 Controls Inflammatory Responses Against External Cues Other than LPS

CD14 participates in other TLR-dependent responses, in addition to those related to TLR4. Indeed, CD14 participates to the signaling downstream of TLR2, TLR3 (Table 1), and TLR9 [126], although the latter has been recently questioned [127].

TLR2 is involved in the recognition of both gram-negative and gram-positive bacteria, mycoplasma, and yeast. CD14 works as a co-receptor capturing ligands for

Table 1 CD14 ligands. POPG: 1-palmitoyl-2-oleoyl-phosphatidylglycerol, S100A9: S100 calcium-binding protein A9, SP-R210: surfactant protein receptor 210, *oxPAPC* oxidized 1-palmitoyl-2-arachidonoyl-sn-glycero-3-phosphocholine, *mmLDL* minimally oxidized low-density lipoprotein, *oxLDL* oxidized low-density lipoprotein, *AD* Alzheimer's disease, *ALS* amyotrophic lateral sclerosis, *FTLD* fronto-temporal lobar degeneration, *CTE* chronic traumatic encephalopathy

Ligand	Source(s)	CD14 interactors and effects	References
Exogenous stressors – Microbial products			
Lipopolysaccharide (LPS)	Gram-negative bacteria	HMGB1, LBP, TLR4-MD2, endocytosis, calcium flux, inflammatory effects	[12, 14, 15, 36, 37, 41, 42, 45, 50, 62–64]
Triacylated lipoproteins	Bacteria	TLR2/1, inflammatory effects	[65–68]
Diacylated lipoproteins	Bacteria	CD36, TLR2/6, inflammatory effects	[68–71]
Peptidoglycan (PGN)	Bacteria	TLR2, inflammatory effects	[72–74]
Lipoteichoic acid (LTA)	Gram-positive bacteria	TLR2/6, inflammatory effects	[69, 75, 76]
Curli fibers	Enterobacteria	TLR2/1, inflammatory effects	[77]
Lipomannan and lipoarabinomannan	Mycobacteria	TLR2/1, inflammatory effects	[21, 69, 78]
Pertussis toxin	*Bordetella pertussis*	Cell adhesion, HIV-1 infection inhibition	[79, 80]
Exogenous stressors – Viral structures			
dsRNA, polyinosine-polycytidylic acid (pIpC)	Virus, synthetic	Regulation of TLR3-dependent responses, inflammatory effects	[81]
Cytomegalovirus	Virus	TLR2, inflammatory effects	[82]
Respiratory syncytial virus	Virus	TLR4, inflammatory effects	[83]
Influenza A virus	Virus	Inflammatory effects	[84]
Hepatitis B virus surface antigen (HBsAg)	Virus	TLR4, inflammatory effects	[85, 86]
Prions and vegetable-derived lectins			
Prions	Chandler and Obihiro strains	Inflammatory effects	[87, 88]
ArtinM	*Artocarpus heterophyllus*	TLR2, inflammatory effects	[89]
Endogenous stressors			
Phosphatidylinositol	Host/Bacteria	Inhibition of TLR4-dependent effects	[90–92]

(continued)

Table 1 (continued)

Ligand	Source(s)	CD14 interactors and effects	References
POPG	Host (lung surfactant)/ Bacteria	Inhibition of TLR4-dependent effects	[92, 93]
oxPAPC	Host	Endocytosis, context-dependent inflammatory or anti-inflammatory effects	[94–96]
mmLDL	Host	TLR4-MD2, cytoskeleton remodeling, efferocytosis inhibition and enhancement of oxLDL uptake	[97]
Apoptotic cells	Host	Efferocytosis	[98–102]
Heat shock family proteins (HSP60, HSP70, HSPA5)	Host	TLR4, TLR2, context-dependent inflammatory or anti-inflammatory effects	[103–106]
Transactive response DNA-binding protein-43 (TDP-43)	Host (ALS, FTLD, AD and CTE patients)	Microglia activation, inflammatory effects	[107]
Amyloid-β	Host (neocortex in the brain of AD patients)	TLR4, TLR2, microglia activation, phagocytosis	[108–112]
Surfactant protein (SP)-A/C/D	Host (lung)	Interactions with LPS signaling, SP-R210	[113–116]
Lactoferrin	Host	Anti-inflammatory effects	[117]
αS1-casein	Host, milk	TLR4-MD2, inflammatory effects	[118]
Versican	Host (extracellular matrix)	TLR2/6, inflammatory effects, enhances metastatic growth	[119]
Biglycan	Host (extracellular matrix)	TLR4 or TLR2, inflammatory effects	[120, 121]
S100A9	Host	TLR4, endocytosis and inflammatory effects	[122]

two distinct TLR2 heterocomplexes: (1) CD14 is implicated in the presentation of triacylated lipoproteins [65–68], lipomannans [21, 69, 78], and curli fibers to TLR2-TLR1 (TLR2/1) [77]; and (2) CD14, with the help of CD36, facilitates the binding to diacylated lipoprotein [68–71], lipoteichoic acid [69, 71, 75], and peptidoglycan to TLR2-TLR6 (TLR2/6) [72–74].

TLR3, an intracellular sensor located in the endosomal compartment, plays a key role in viral infections, recognizing double strand (ds) RNA. CD14 works as a shuttle for extracellular polyinosine-polycytidylic acid (pIpC), inducing its uptake and the physical interaction with TLR3 in lysosomes, where TLR3 starts the signaling cascades that yield to cell activation [81]. CD14 also plays an additional role in antiviral responses, recognizing lipid and glycosylated moieties of viral membrane envelopes and capsids and, thus, leading to the activation of TLR4- and/or TLR2-dependent responses [82–86].

3.2.2 CD14 Mediates the Responses to Endogenous Lipidic Stress Signals

CD14 interacts with host-derived moieties (Table 1) that are produced or accumulated during stress conditions, such as tissue damage (associated or not with pathogenic infections), metabolic disorders, protein aggregation diseases, and neoplastic processes.

An important example is atherosclerosis, a disease characterized by the deposition of fatty material, such as low-density lipoprotein (LDL), in the inner walls of blood vessels. In this context, phagocytes in the intima are responsible for the development of a detrimental chronic inflammatory response. CD14 controls several functions of these cells, regulating the binding to fatty structures and moieties. Indeed, CD14 controls the activity of minimally modified LDL (mmLDL) [97], produced by a mild oxidative process in the intima. mmLDL triggers via CD14 a TLR4-MD2-depedent inflammatory response sustained by SYK and the downstream activation of Vav Guanine Nucleotide Exchange Factor 1 (VAV1), PLCγ, and c-Jun N-terminal kinases (JNKs). In turn, these pathways induce the production of reactive oxygen species (ROS), cytokine secretion, and cytoskeleton rearrangements [97, 128–130], thereby supporting pro-inflammatory macrophage activation and pro-atherogenic activities (Fig. 3).

Efferocytosis is another process that can be regulated by CD14 [98–102]. During atherosclerosis progression, the unbalance between dead cell accumulation and their removal leads to the formation of a necrotic core in the atheroma, which highly increases the risk of plaque rupture and the formation of thrombi [131]. The N-terminal domain of CD14 has been shown to mediate the binding of apoptotic bodies, probably through the interaction with modified lipidic moieties, and to induce their internalization in macrophages, thus favoring the clearance of dead cells [102] and participating to the control of the evolution of this pathology (Fig. 3).

Interestingly, both mmLDL and dying cells contain oxidized lipids, such as oxidized phospholipid 1-palmitoyl-2-arachidonoyl-sn-glycero-3-phosphorylcholine (oxPAPC), which activates macrophages in CD14-dependent and independent manners [94, 132, 133]. oxPAPC can be internalized by CD14 similarly to LPS (see 2.1.2). Once internalized in a CD14-dependent manner, oxPAPC triggers the activation of the inflammasome by binding to caspase-11 [94]. In contrast to LPS and other inflammasome-activating canonical stimuli, oxPAPC induces the cleavage of IL-1 without inducing pyroptosis [94, 132]. In addition, oxPAPC governs a CD14-independent metabolic remodeling necessary to produce regulatory metabolites, such as oxaloacetate, that potentiate the inflammatory response [133] (Fig. 3). In association with subclinical endotoxemia [134], these two mechanisms sustain and boost IL-1β production from long-lived macrophages, eliciting a prolonged and detrimental hyper-inflammation that sustains atherosclerosis development [133].

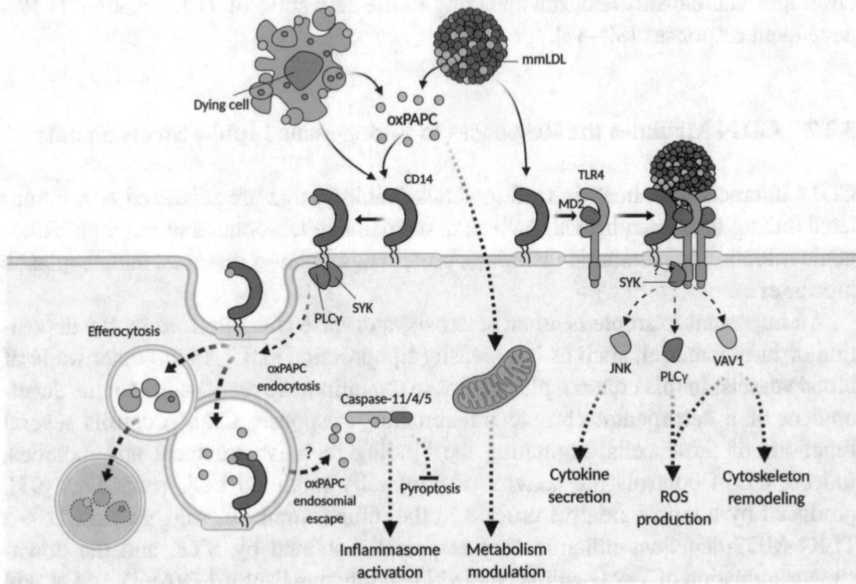

Fig. 3 CD14 controls endogenous lipidic stress signals. During atherosclerosis, fat-containing material is accumulated in the intima of blood vessels where phagocytes start an inflammatory response. mmLDL is recognized by the CD14-TLR4-MD2 complex that coordinates SYK-dependent activities via JNK, PLCγ, and VAV1 activation, inducing cytokine secretion, ROS production, and cytoskeleton remodeling. Dying cells are cleared by efferocytosis induced by CD14, controlling atheroma composition. The accumulation of oxidized lipids, such as oxPAPC, in the atherosclerotic plaque can trigger CD14-independent and CD14-dependent responses. oxPAPC modifies the cellular metabolism and induces the accumulation of specific metabolites that boost IL-1β production. CD14 mediates oxPAPC internalization. oxPAPC in the cytosol binds caspase-11/4/5, activating inflammasome-dependent functions without inducing pyroptosis

3.2.3 CD14 Mediates the Responses to Endogenous Protein Stress Signals

In addition to host-derived lipidic structures and mirroring its ability to bind a large variety of exogenous ligands, CD14 also controls the inflammatory responses triggered by specific stress proteins (Table 1). Heat-shock family proteins (HSFPs) are released in the extracellular milieu as a result of tissue damage or stress conditions [135]. HSFPs can activate TLR4 and TLR2-depedent responses via CD14 [103–105]. HSFPs, such as HSPA5 (also known as GRP-78), can also trigger TLR4-MD2 internalization via CD14, creating a functional TLR4 deficiency that dampens the inflammatory response elicited by LPS [106]. CD14 ligands can also be produced during lung infections: the secretion of surfactant proteins can be enhanced and can modulate the activity of the complex formed by CD14-TLR4-MD2 [113–116]. Another source of moieties able to bind CD14 is represented by protein aggregates found in many neurodegenerative diseases. Transactive response DNA-binding protein-43 (TDP-43) [136] is present in tissue inclusions of amyotrophic lateral sclerosis (ALS), fronto-temporal lobar degeneration (FTLD) [137], Alzheimer's disease

(AD) [138], and chronic traumatic encephalopathy [139], and it has been shown that microglial CD14 is necessary for TDP-43-induced inflammation and neurotoxicity [107]. CD14 can also interact with amyloid fibers, hallmarks of AD, inducing their cellular internalization and the triggering of local inflammation via TLR4 or TLR2 [108–112]. Notably, a similar recognition might occur also for prion fibers [87, 88], suggesting that CD14 indiscriminately recognizes large extracellular protein aggregates. Lastly, CD14 has been shown to participate to the recognition of moieties associated with tumor growth. Cancer cells can over-produce extracellular matrix (ECM) proteins or secrete specific proteases that modified the pre-existed ECM, creating permissive conditions for tumor migration and spreading. In this context, CD14 on phagocytes can recognize specific ECM elements, such as versican [119], and activate inflammatory responses that modulate the growth of metastasis and tumor progression.

4 Concluding Remarks

CD14 is a key player during LPS-induced inflammatory responses, controlling critical aspects of TLR4 biology and its complementary processes. CD14 also works as a versatile "molecular sentinel" that is able to recognize a plethora of tissue stressors and to initiate and/or coordinate both inflammatory (i.e., pro-inflammatory cytokine production) and anti-inflammatory (i.e., clearance of apoptotic cells) responses. Its wide body distribution and its capacity to control multiple and different ligand-receptor interactions, both in immune and non-immune cells, make it an important player in several homeostatic and pathological programs induced by microbial and host-derived threats. In fact, CD14 has been associated with several diseases, such as sepsis, autoimmune diseases, metabolic syndrome, atherosclerosis, neurodegenerative conditions, and cancer. Therefore, understanding the *context-dependent* activity of CD14, in relation to ligands, receptors, cell types and extracellular milieu, might provide novel insights for the development of new clinical treatments and might expand our knowledge about the interplay between the immune system and other body systems.

References

1. Haziot A, Chen S, Ferrero E, Low MG, Silber R, Goyert SM. The monocyte differentiation antigen, CD14, is anchored to the cell membrane by a phosphatidylinositol linkage. J Immunol. 1988;141:547–52.
2. Schroeder R, London E, Brown D. Interactions between saturated acyl chains confer detergent resistance on lipids and glycosylphosphatidylinositol (GPI)-anchored proteins: GPI-anchored proteins in liposomes and cells show similar behavior. Proc Natl Acad Sci. 1994;91:12130–4. https://doi.org/10.1073/pnas.91.25.12130.
3. Bazil V, Strominger JL. Shedding as a mechanism of down-modulation of CD14 on stimulated human monocytes. J Immunol (Baltim, Md: 1950). 1991;147:1567–74.

4. Bufler P, Stiegler G, Schuchmann M, Hess S, Krüger C, Stelter F, Eckerskorn C, Schütt C, Engelmann H. Soluble lipopolysaccharide receptor (CD14) is released via two different mechanisms from human monocytes and CD14 transfectants. Eur J Immunol. 1995;25:604–10. https://doi.org/10.1002/eji.1830250244.
5. Metz CN, Brunner G, Choi-Muira NH, Nguyen H, Gabrilove J, Caras IW, Altszuler N, Rifkin DB, Wilson EL, Davitz MA. Release of GPI-anchored membrane proteins by a cell-associated GPI-specific phospholipase D. EMBO J. 1994;13:1741–51.
6. Du X, Low MG. Down-regulation of glycosylphosphatidylinositol-specific phospholipase D induced by lipopolysaccharide and oxidative stress in the murine monocyte- macrophage cell line RAW 264.7. Infect Immun. 2001;69:3214–23. https://doi.org/10.1128/IAI.69.5.3214-3223.2001.
7. Kelley SL, Lukk T, Nair SK, Tapping RI. The crystal structure of human soluble CD14 reveals a bent solenoid with a hydrophobic amino-terminal pocket. J Immunol. 2013;190:1304–11. https://doi.org/10.4049/jimmunol.1202446.
8. UniProt Consortium. UniProt: a worldwide hub of protein knowledge. Nucleic Acids Res. 2019;47:D506–15. https://doi.org/10.1093/nar/gky1049.
9. Kim J-I, Lee CJ, Jin MS, Lee C-H, Paik S-G, Lee H, Lee J-O. Crystal structure of CD14 and its implications for lipopolysaccharide signaling. J Biol Chem. 2005;280:11347–51. https://doi.org/10.1074/jbc.M414607200.
10. Uhlén M, Fagerberg L, Hallström BM, Lindskog C, Oksvold P, Mardinoglu A, Sivertsson Å, Kampf C, Sjöstedt E, Asplund A, Olsson I, Edlund K, Lundberg E, Navani S, Szigyarto CA-K, Odeberg J, Djureinovic D, Takanen JO, Hober S, Alm T, Edqvist P-H, Berling H, Tegel H, Mulder J, Rockberg J, Nilsson P, Schwenk JM, Hamsten M, von Feilitzen K, Forsberg M, Persson L, Johansson F, Zwahlen M, von Heijne G, Nielsen J, Pontén F. Tissue-based map of the human proteome. Science. 2015;347:1260419. https://doi.org/10.1126/science.1260419.
11. Tissue expression of CD14 – Summary – The Human Protein Atlas. https://www.proteinatlas.org/ENSG00000170458-CD14/tissue. Accessed 1 Dec 2019.
12. Wright SD, Ramos RA, Tobias PS, Ulevitch RJ, Mathison JC. CD14, a receptor for complexes of lipopolysaccharide (LPS) and LPS binding protein. Science. 1990;249:1431–3. https://doi.org/10.1126/science.1698311.
13. Stelter F, Bernheiden M, Menzel R, Jack RS, Witt S, Fan X, Pfister M, Schütt C. Mutation of amino acids 39-44 of human CD14 abrogates binding of lipopolysaccharide and Escherichia coli. Eur J Biochem. 1997;243:100–9. https://doi.org/10.1111/j.1432-1033.1997.00100.x.
14. Tan Y, Zanoni I, Cullen TW, Goodman AL, Kagan JC. Mechanisms of toll-like receptor 4 endocytosis reveal a common immune-evasion strategy used by pathogenic and commensal bacteria. Immunity. 2015;43:909–22. https://doi.org/10.1016/j.immuni.2015.10.008.
15. Ryu J-K, Kim SJ, Rah S-H, Kang JI, Jung HE, Lee D, Lee HK, Lee J-O, Park BS, Yoon T-Y, Kim HM. Reconstruction of LPS transfer Cascade reveals structural determinants within LBP, CD14, and TLR4-MD2 for efficient LPS recognition and transfer. Immunity. 2017;46:38–50. https://doi.org/10.1016/j.immuni.2016.11.007.
16. Huber RG, Berglund NA, Kargas V, Marzinek JK, Holdbrook DA, Khalid S, Piggot TJ, Schmidtchen A, Bond PJ. A thermodynamic funnel drives bacterial lipopolysaccharide transfer in the TLR4 pathway. Structure. 2018;26:1151–1161.e4. https://doi.org/10.1016/j.str.2018.04.007.
17. Goyert SM, Ferrero EM, Seremetis SV, Winchester RJ, Silver J, Mattison AC. Biochemistry and expression of myelomonocytic antigens. J Immunol (Baltim, Md: 1950). 1986;137:3909–14.
18. Jersmann HP. Time to abandon dogma: CD14 is expressed by non-myeloid lineage cells. Immunol Cell Biol. 2005;83:462–7. https://doi.org/10.1111/j.1440-1711.2005.01370.x.
19. Park H-J, Lee W-Y, Park C, Hong K, Song H. CD14 is a unique membrane marker of porcine spermatogonial stem cells, regulating their differentiation. Sci Rep. 2019;9:1–11. https://doi.org/10.1038/s41598-019-46000-6.
20. Lien E, Aukrust P, Sundan A, Müller F, Frøland SS, Espevik T. Elevated levels of serum-soluble CD14 in human immunodeficiency virus type 1 (HIV-1) infection: correlation to disease progression and clinical events. Blood. 1998;92:2084–92.

21. Pugin J, Heumann ID, Tomasz A, Kravchenko VV, Akamatsu Y, Nishijima M, Glauser MP, Tobias PS, Ulevitch RJ. CD14 is a pattern recognition receptor. Immunity. 1994;1:509–16. https://doi.org/10.1016/1074-7613(94)90093-0.
22. Medzhitov R, Preston-Hurlburt P, Janeway CA. A human homologue of the Drosophila toll protein signals activation of adaptive immunity. Nature. 1997;388:394–7. https://doi.org/10.1038/41131.
23. Poltorak A, He X, Smirnova I, Liu M-Y, Huffel CV, Du X, Birdwell D, Alejos E, Silva M, Galanos C, Freudenberg M, Ricciardi-Castagnoli P, Layton B, Beutler B. Defective LPS signaling in C3H/HeJ and C57BL/10ScCr mice: mutations in Tlr4 gene. Science. 1998;282:2085–8. https://doi.org/10.1126/science.282.5396.2085.
24. Haziot A, Ferrero E, Köntgen F, Hijiya N, Yamamoto S, Silver J, Stewart CL, Goyert SM. Resistance to endotoxin shock and reduced dissemination of gram-negative Bacteria in CD14-deficient mice. Immunity. 1996;4:407–14. https://doi.org/10.1016/S1074-7613(00)80254-X.
25. Wiersinga WJ, de Vos AF, Wieland CW, Leendertse M, Roelofs JJTH, van der Poll T. CD14 impairs host defense against gram-negative sepsis caused by Burkholderia pseudomallei in mice. J Infect Dis. 2008;198:1388–97. https://doi.org/10.1086/592220.
26. Namath A, Patterson AJ. Genetic polymorphisms in sepsis. Crit Care Nurs Clin North Am. 2011;23:181–202. https://doi.org/10.1016/j.ccell.2010.12.011.
27. Mansur A, Liese B, Steinau M, Ghadimi M, Bergmann I, Tzvetkov M, Popov AF, Beissbarth T, Bauer M, Hinz J. The CD14 rs 2569190 TT genotype is associated with an improved 30-day survival in patients with Sepsis: a prospective observational cohort study. PLoS One. 2015;10:e0127761. https://doi.org/10.1371/journal.pone.0127761.
28. Jiménez-Sousa MÁ, Liu P, Medrano LM, Fernández-Rodríguez A, Almansa R, Gómez-Sánchez E, Rico L, Lorenzo M, Fadrique A, Tamayo E, Resino S. Association of CD14 rs2569190 polymorphism with mortality in shock septic patients who underwent major cardiac or abdominal surgery: a retrospective study. Sci Rep. 2018;8:2698. https://doi.org/10.1038/s41598-018-20766-7.
29. de Aguiar BB, Girardi I, Paskulin DD, de Franca E, Dornelles C, Dias FS, Bonorino C, Alho CS. CD14 expression in the first 24h of sepsis: effect of -260C>T CD14 SNP. Immunol Investig. 2008;37:752–69. https://doi.org/10.1080/08820130802403242.
30. Zhang A, Yue C, Gu W, Du J, Wang H, Jiang J. Association between CD14 promoter -159C/T polymorphism and the risk of sepsis and mortality: a systematic review and meta-analysis. PLoS One. 2013;8:e71237. https://doi.org/10.1371/journal.pone.0071237.
31. Lin J, Yao Y, Yu Y, Chai J, Huang Z, Dong N, Sheng Z. Effects of CD14-159 C/T polymorphism on CD14 expression and the balance between proinflammatory and anti-inflammatory cytokines in whole blood culture. Shock Augusta Ga. 2007;28:148–53. https://doi.org/10.1097/SHK.0b013e3180341d35.
32. Gu W, Dong H, Jiang D-P, Zhou J, Du D-Y, Gao J-M, Yao Y-Z, Zhang L-Y, Wen A-Q, Liu Q, Wang Z-G, Jiang J-X. Functional significance of CD14 promoter polymorphisms and their clinical relevance in a Chinese Han population*. Crit Care Med. 2008;36:2274–80. https://doi.org/10.1097/CCM.0b013e318180b1ed.
33. Beveridge TJ. Structures of gram-negative cell walls and their derived membrane vesicles. J Bacteriol. 1999;181:4725–33.
34. Vanaja SK, Russo AJ, Behl B, Banerjee I, Yankova M, Deshmukh SD, Rathinam VAK. Bacterial outer membrane vesicles mediate cytosolic localization of LPS and Caspase-11 activation. Cell. 2016;165:1106–19. https://doi.org/10.1016/j.cell.2016.04.015.
35. Deng M, Tang Y, Li W, Wang X, Zhang R, Zhang X, Zhao X, Liu J, Tang C, Liu Z, Huang Y, Peng H, Xiao L, Tang D, Scott MJ, Wang Q, Liu J, Xiao X, Watkins S, Li J, Yang H, Wang H, Chen F, Tracey KJ, Billiar TR, Lu B. The endotoxin delivery protein HMGB1 mediates Caspase-11-dependent lethality in sepsis. Immunity. 2018;49:740–753.e7. https://doi.org/10.1016/j.immuni.2018.08.016.

36. Youn JH, Oh YJ, Kim ES, Choi JE, Shin J-S. High mobility group box 1 protein binding to lipopolysaccharide facilitates transfer of lipopolysaccharide to CD14 and enhances lipopolysaccharide-mediated TNF-α production in human monocytes. J Immunol. 2008;180:5067–74. https://doi.org/10.4049/jimmunol.180.7.5067.
37. Gioannini TL, Teghanemt A, Zhang D, Coussens NP, Dockstader W, Ramaswamy S, Weiss JP. Isolation of an endotoxin–MD-2 complex that produces Toll-like receptor 4-dependent cell activation at picomolar concentrations. Proc Natl Acad Sci. 2004;101:4186–91. https://doi.org/10.1073/pnas.0306906101.
38. Triantafilou M, Miyake K, Golenbock DT, Triantafilou K. Mediators of innate immune recognition of bacteria concentrate in lipid rafts and facilitate lipopolysaccharide-induced cell activation. J Cell Sci. 2002;115:2603–11.
39. Kagan JC, Medzhitov R. Phosphoinositide-mediated adaptor recruitment controls toll-like receptor signaling. Cell. 2006;125:943–55. https://doi.org/10.1016/j.cell.2006.03.047.
40. Kagan JC, Su T, Horng T, Chow A, Akira S, Medzhitov R. TRAM couples endocytosis of toll-like receptor 4 to the induction of interferon-beta. Nat Immunol. 2008;9:361–8. https://doi.org/10.1038/ni1569.
41. Jiang Z, Georgel P, Du X, Shamel L, Sovath S, Mudd S, Huber M, Kalis C, Keck S, Galanos C, Freudenberg M, Beutler B. CD14 is required for MyD88-independent LPS signaling. Nat Immunol. 2005;6:565–70. https://doi.org/10.1038/ni1207.
42. Zanoni I, Ostuni R, Marek LR, Barresi S, Barbalat R, Barton GM, Granucci F, Kagan JC. CD14 controls the LPS-induced endocytosis of toll-like receptor 4. Cell. 2011;147:868–80. https://doi.org/10.1016/j.cell.2011.09.051.
43. Botelho RJ, Teruel M, Dierckman R, Anderson R, Wells A, York JD, Meyer T, Grinstein S. Localized biphasic changes in phosphatidylinositol-4,5-bisphosphate at sites of phagocytosis. J Cell Biol. 2000;151:1353–68. https://doi.org/10.1083/jcb.151.7.1353.
44. Chiang C-Y, Veckman V, Limmer K, David M. Phospholipase Cγ-2 and intracellular calcium are required for lipopolysaccharide-induced Toll-like receptor 4 (TLR4) endocytosis and interferon regulatory factor 3 (IRF3) activation. J Biol Chem. 2012;287:3704–9. https://doi.org/10.1074/jbc.C111.328559.
45. Schappe MS, Szteyn K, Stremska ME, Mendu SK, Downs TK, Seegren PV, Mahoney MA, Dixit S, Krupa JK, Stipes EJ, Rogers JS, Adamson SE, Leitinger N, Desai BN. Chanzyme TRPM7 mediates the Ca2+ influx essential for lipopolysaccharide-induced toll-like receptor 4 endocytosis and macrophage activation. Immunity. 2018;48:59–74.e5. https://doi.org/10.1016/j.immuni.2017.11.026.
46. Kovarik P, Castiglia V, Ivin M, Ebner F. Type I interferons in bacterial infections: a balancing act. Front Immunol. 2016;7 https://doi.org/10.3389/fimmu.2016.00652.
47. Souza DPD, Achuthan A, Lee MKS, Binger KJ, Lee M-C, Davidson S, Tull DL, McConville MJ, Cook AD, Murphy AJ, Hamilton JA, Fleetwood AJ. Autocrine IFN-I inhibits isocitrate dehydrogenase in the TCA cycle of LPS-stimulated macrophages. J Clin Invest. 2019;129:4239–44. https://doi.org/10.1172/JCI127597.
48. Mills EL, Ryan DG, Prag HA, Dikovskaya D, Menon D, Zaslona Z, Jedrychowski MP, Costa ASH, Higgins M, Hams E, Szpyt J, Runtsch MC, King MS, McGouran JF, Fischer R, Kessler BM, McGettrick AF, Hughes MM, Carroll RG, Booty LM, Knatko EV, Meakin PJ, Ashford MLJ, Modis LK, Brunori G, Sévin DC, Fallon PG, Caldwell ST, Kunji ERS, Chouchani ET, Frezza C, Dinkova-Kostova AT, Hartley RC, Murphy MP, O'Neill LA. Itaconate is an anti-inflammatory metabolite that activates Nrf2 via alkylation of KEAP1. Nature. 2018;556:113–7. https://doi.org/10.1038/nature25986.
49. Postat J, Olekhnovitch R, Lemaître F, Bousso P. A metabolism-based quorum sensing mechanism contributes to termination of inflammatory responses. Immunity. 2018;49:654–665.e5. https://doi.org/10.1016/j.immuni.2018.07.014.
50. Zanoni I, Ostuni R, Capuano G, Collini M, Caccia M, Ronchi AE, Rocchetti M, Mingozzi F, Foti M, Chirico G, Costa B, Zaza A, Ricciardi-Castagnoli P, Granucci F. CD14 regulates the dendritic cell life cycle after LPS exposure through NFAT activation. Nature. 2009;460:264–8. https://doi.org/10.1038/nature08118.

51. Dagvadorj J, Shimada K, Chen S, Jones HD, Tumurkhuu G, Zhang W, Wawrowsky KA, Crother TR, Arditi M. Lipopolysaccharide induces alveolar macrophage necrosis via CD14 and the P2X7 receptor leading to interleukin-1α release. Immunity. 2015;42:640–53. https://doi.org/10.1016/j.immuni.2015.03.007.
52. Zanoni I, Ostuni R, Barresi S, Gioia MD, Broggi A, Costa B, Marzi R, Granucci F. CD14 and NFAT mediate lipopolysaccharide-induced skin edema formation in mice. J Clin Invest. 2012;122:1747–57. https://doi.org/10.1172/JCI60688.
53. Zanoni I, Spreafico R, Bodio C, Di Gioia M, Cigni C, Broggi A, Gorletta T, Caccia M, Chirico G, Sironi L, Collini M, Colombo MP, Garbi N, Granucci F. IL-15 cis presentation is required for optimal NK cell activation in lipopolysaccharide-mediated inflammatory conditions. Cell Rep. 2013;4:1235–49. https://doi.org/10.1016/j.celrep.2013.08.021.
54. Wuest SC, Edwan JH, Martin JF, Han S, Perry JSA, Cartagena CM, Matsuura E, Maric D, Waldmann TA, Bielekova B. A role for interleukin-2 trans-presentation in dendritic cell-mediated T cell activation in humans, as revealed by daclizumab therapy. Nat Med. 2011;17:604–9. https://doi.org/10.1038/nm.2365.
55. Zelante T, Wong AYW, Ping TJ, Chen J, Sumatoh HR, Viganò E, Hong Bing Y, Lee B, Zolezzi F, Fric J, Newell EW, Mortellaro A, Poidinger M, Puccetti P, Ricciardi-Castagnoli P. CD103(+) dendritic cells control Th17 cell function in the lung. Cell Rep. 2015;12:1789–801. https://doi.org/10.1016/j.celrep.2015.08.030.
56. Kulhankova K, Rouse T, Nasr ME, Field EH. Dendritic cells control CD4+CD25+ Treg cell suppressor function in vitro through juxtacrine delivery of IL-2. PLoS One. 2012;7:e43609. https://doi.org/10.1371/journal.pone.0043609.
57. Sgouroudis E, Kornete M, Piccirillo CA. IL-2 production by dendritic cells promotes Foxp3(+) regulatory T-cell expansion in autoimmune-resistant NOD congenic mice. Autoimmunity. 2011;44:406–14. https://doi.org/10.3109/08916934.2010.536795.
58. Mencarelli A, Khameneh HJ, Fric J, Vacca M, El Daker S, Janela B, Tang JP, Nabti S, Balachander A, Lim TS, Ginhoux F, Ricciardi-Castagnoli P, Mortellaro A. Calcineurin-mediated IL-2 production by CD11chighMHCII+ myeloid cells is crucial for intestinal immune homeostasis. Nat Commun. 2018;9 https://doi.org/10.1038/s41467-018-03495-3.
59. Li Q-X, Ke N, Sundaram R, Wong-Staal F. NR4A1, 2, 3--an orphan nuclear hormone receptor family involved in cell apoptosis and carcinogenesis. Histol Histopathol. 2006;21:533–40. https://doi.org/10.14670/HH-21.533.
60. Zanoni I, Granucci F. The regulatory role of dendritic cells in the induction and maintenance of T-cell tolerance. Autoimmunity. 2011;44:23–32. https://doi.org/10.3109/08916931003782148.
61. Denlinger LC, Fisette PL, Sommer JA, Watters JJ, Prabhu U, Dubyak GR, Proctor RA, Bertics PJ. Cutting edge: the nucleotide receptor P2X7 contains multiple protein- and lipid-interaction motifs including a potential binding site for bacterial lipopolysaccharide. J Immunol (Baltim, Md: 1950). 2001;167:1871–6. https://doi.org/10.4049/jimmunol.167.4.1871.
62. da Silva Correia J, Soldau K, Christen U, Tobias PS, Ulevitch RJ. Lipopolysaccharide is in close proximity to each of the proteins in its membrane receptor complex TRANSFER FROM CD14 TO TLR4 AND MD-2. J Biol Chem. 2001;276:21129–35. https://doi.org/10.1074/jbc.M009164200.
63. Moore KJ, Andersson LP, Ingalls RR, Monks BG, Li R, Arnaout MA, Golenbock DT, Freeman MW. Divergent response to LPS and Bacteria in CD14-deficient murine macrophages. J Immunol. 2000;165:4272–80. https://doi.org/10.4049/jimmunol.165.8.4272.
64. Kim S, Kim SY, Pribis JP, Lotze M, Mollen KP, Shapiro R, Loughran P, Scott MJ, Billiar TR. Signaling of high mobility group box 1 (HMGB1) through toll-like receptor 4 in macrophages requires CD14. Mol Med (Cambridge, MA). 2013;19:88–98. https://doi.org/10.2119/molmed.2012.00306.
65. Nakata T, Yasuda M, Fujita M, Kataoka H, Kiura K, Sano H, Shibata K. CD14 directly binds to triacylated lipopeptides and facilitates recognition of the lipopeptides by the receptor complex of Toll-like receptors 2 and 1 without binding to the complex. Cell Microbiol. 2006;8:1899–909. https://doi.org/10.1111/j.1462-5822.2006.00756.x.

66. Manukyan M, Triantafilou K, Triantafilou M, Mackie A, Nilsen N, Espevik T, Wiesmüller K-H, Ulmer AJ, Heine H. Binding of lipopeptide to CD14 induces physical proximity of CD14, TLR2 and TLR1. Eur J Immunol. 2005;35:911–21. https://doi.org/10.1002/eji.200425336.
67. Ranoa DRE, Kelley SL, Tapping RI. Human lipopolysaccharide-binding protein (LBP) and CD14 independently deliver Triacylated lipoproteins to toll-like receptor 1 (TLR1) and TLR2 and enhance formation of the ternary signaling complex. J Biol Chem. 2013;288:9729–41. https://doi.org/10.1074/jbc.M113.453266.
68. Triantafilou M, Gamper FGJ, Haston RM, Mouratis MA, Morath S, Hartung T, Triantafilou K. Membrane sorting of toll-like receptor (TLR)-2/6 and TLR2/1 heterodimers at the cell surface determines heterotypic associations with CD36 and intracellular targeting. J Biol Chem. 2006;281:31002–11. https://doi.org/10.1074/jbc.M602794200.
69. Jimenez-Dalmaroni MJ, Xiao N, Corper AL, Verdino P, Ainge GD, Larsen DS, Painter GF, Rudd PM, Dwek RA, Hoebe K, Beutler B, Wilson IA. Soluble CD36 ectodomain binds negatively charged diacylglycerol ligands and acts as a co-receptor for TLR2. PLoS One. 2009;4:e7411. https://doi.org/10.1371/journal.pone.0007411.
70. Mae M, Iyori M, Yasuda M, Shamsul HM, Kataoka H, Kiura K, Hasebe A, Totsuka Y, Shibata K-I. The diacylated lipopeptide FSL-1 enhances phagocytosis of bacteria by macrophages through a toll-like receptor 2-mediated signalling pathway. FEMS Immunol Med Microbiol. 2007;49:398–409. https://doi.org/10.1111/j.1574-695X.2007.00218.x.
71. Schröder NWJ, Heine H, Alexander C, Manukyan M, Eckert J, Hamann L, Göbel UB, Schumann RR. Lipopolysaccharide binding protein binds to Triacylated and Diacylated Lipopeptides and mediates innate immune responses. J Immunol. 2004;173:2683–91. https://doi.org/10.4049/jimmunol.173.4.2683.
72. Gupta D, Kirkland TN, Viriyakosol S, Dziarski R. CD14 is a cell-activating receptor for bacterial peptidoglycan. J Biol Chem. 1996;271:23310–6. https://doi.org/10.1074/jbc.271.38.23310.
73. Dziarski R, Tapping RI, Tobias PS. Binding of bacterial peptidoglycan to CD14. J Biol Chem. 1998;273:8680–90. https://doi.org/10.1074/jbc.273.15.8680.
74. Muhvic D, El-Samalouti V, Flad H-D, Radoševic-Stašic B, Rukavina D (2001) The involvement of CD14 in the activation of human monocytes by peptidoglycan monomers. In: Mediators inflamm. https://www.hindawi.com/journals/mi/2001/150296/abs/. Accessed 2 Dec 2019.
75. Cleveland MG, Gorham JD, Murphy TL, Tuomanen E, Murphy KM. Lipoteichoic acid preparations of gram-positive bacteria induce interleukin-12 through a CD14-dependent pathway. Infect Immun. 1996;64:1906–12.
76. Schröder NWJ, Morath S, Alexander C, Hamann L, Hartung T, Zähringer U, Göbel UB, Weber JR, Schumann RR. Lipoteichoic acid (LTA) of Streptococcus pneumoniaeand Staphylococcus aureus activates immune cells via toll-like receptor (TLR)-2, lipopolysaccharide-binding protein (LBP), and CD14, whereas TLR-4 and MD-2 are not involved. J Biol Chem. 2003;278:15587–94. https://doi.org/10.1074/jbc.M212829200.
77. Rapsinski GJ, Newman TN, Oppong GO, van Putten JPM, Tükel Ç. CD14 protein acts as an adaptor molecule for the immune recognition of salmonella curli fibers. J Biol Chem. 2013;288:14178–88. https://doi.org/10.1074/jbc.M112.447060.
78. Elass E, Coddeville B, Guérardel Y, Kremer L, Maes E, Mazurier J, Legrand D. Identification by surface plasmon resonance of the mycobacterial lipomannan and lipoarabinomannan domains involved in binding to CD14 and LPS-binding protein. FEBS Lett. 2007;581:1383–90. https://doi.org/10.1016/j.febslet.2007.02.056.
79. Li H, Wong WSF. Mechanisms of pertussis toxin-induced myelomonocytic cell adhesion: role of CD14 and urokinase receptor. Immunology. 2000;100:502–9. https://doi.org/10.1046/j.1365-2567.2000.00064.x.
80. Hu Q, Younson J, Griffin GE, Kelly C, Shattock RJ. Pertussis toxin and its binding unit inhibit HIV-1 infection of human cervical tissue and macrophages involving a CD14 pathway. J Infect Dis. 2006;194:1547–56. https://doi.org/10.1086/508898.

81. Lee H-K, Dunzendorfer S, Soldau K, Tobias PS. Double-stranded RNA-mediated TLR3 activation is enhanced by CD14. Immunity. 2006;24:153–63. https://doi.org/10.1016/j.immuni.2005.12.012.
82. Compton T, Kurt-Jones EA, Boehme KW, Belko J, Latz E, Golenbock DT, Finberg RW. Human Cytomegalovirus activates inflammatory cytokine responses via CD14 and toll-like receptor 2. J Virol. 2003;77:4588–96. https://doi.org/10.1128/JVI.77.8.4588-4596.2003.
83. Kurt-Jones EA, Popova L, Kwinn L, Haynes LM, Jones LP, Tripp RA, Walsh EE, Freeman MW, Golenbock DT, Anderson LJ, Finberg RW. Pattern recognition receptors TLR4 and CD14 mediate response to respiratory syncytial virus. Nat Immunol. 2000;1:398–401. https://doi.org/10.1038/80833.
84. Pauligk C, Nain M, Reiling N, Gemsa D, Kaufmann A. CD14 is required for influenza a virus-induced cytokine and chemokine production. Immunobiology. 2004;209:3–10. https://doi.org/10.1016/j.imbio.2004.04.002.
85. Vanlandschoot P, Van Houtte F, Roobrouck A, Farhoudi A, Stelter F, Peterson DL, Gomez-Gutierrez J, Gavilanes F, Leroux-Roels G. LPS-binding protein and CD14-dependent attachment of hepatitis B surface antigen to monocytes is determined by the phospholipid moiety of the particles. J Gen Virol. 2002;83:2279–89. https://doi.org/10.1099/0022-1317-83-9-2279.
86. van Montfoort N, van der Aa E, van den Bosch A, Brouwers H, Vanwolleghem T, Janssen HLA, Javanbakht H, Buschow SI, Woltman AM. Hepatitis B virus surface antigen activates myeloid dendritic cells via a soluble CD14-dependent mechanism. J Virol. 2016;90:6187–99. https://doi.org/10.1128/JVI.02903-15.
87. Sakai K, Hasebe R, Takahashi Y, Song C-H, Suzuki A, Yamasaki T, Horiuchi M. Absence of CD14 delays progression of prion diseases accompanied by increased microglial activation. J Virol. 2013;87:13433–45. https://doi.org/10.1128/JVI.02072-13.
88. Hasebe R, Suzuki A, Yamasaki T, Horiuchi M. Temporary upregulation of anti-inflammatory cytokine IL-13 expression in the brains of CD14 deficient mice in the early stage of prion infection. Biochem Biophys Res Commun. 2014;454:125–30. https://doi.org/10.1016/j.bbrc.2014.10.043.
89. da Silva TA, Zorzetto-Fernandes ALV, Cecílio NT, Sardinha-Silva A, Fernandes FF, Roque-Barreira MC. CD14 is critical for TLR2-mediated M1 macrophage activation triggered by N-glycan recognition. Sci Rep. 2017;7:1–14. https://doi.org/10.1038/s41598-017-07397-0.
90. Yu B, Hailman E, Wright SD. Lipopolysaccharide binding protein and soluble CD14 catalyze exchange of phospholipids. J Clin Invest. 1997;99:315–24. https://doi.org/10.1172/JCI119160.
91. Wang P, Kitchens RL, Munford RS. Phosphatidylinositides bind to plasma membrane CD14 and can prevent monocyte activation by bacterial lipopolysaccharide. J Biol Chem. 1998;273:24309–13. https://doi.org/10.1074/jbc.273.38.24309.
92. Kuronuma K, Mitsuzawa H, Takeda K, Nishitani C, Chan ED, Kuroki Y, Nakamura M, Voelker DR. Anionic pulmonary surfactant phospholipids inhibit inflammatory responses from alveolar macrophages and U937 cells by binding the lipopolysaccharide interacting proteins CD14 and MD2. J Biol Chem. 2009;284(38):25488–500. https://doi.org/10.1074/jbc.M109.040832.
93. Hashimoto M, Asai Y, Ogawa T. Treponemal phospholipids inhibit innate immune responses induced by pathogen-associated molecular patterns. J Biol Chem. 2003;278:44205–13. https://doi.org/10.1074/jbc.M306735200.
94. Zanoni I, Tan Y, Di Gioia M, Springstead JR, Kagan JC. By capturing inflammatory lipids released from dying cells, the receptor CD14 induces inflammasome-dependent phagocyte Hyperactivation. Immunity. 2017;47:697–709.e3. https://doi.org/10.1016/j.immuni.2017.09.010.
95. Erridge C, Kennedy S, Spickett CM, Webb DJ. Oxidized phospholipid inhibition of toll-like receptor (TLR) signaling is restricted to TLR2 and TLR4 ROLES FOR CD14, LPS-BINDING PROTEIN, AND MD2 AS TARGETS FOR SPECIFICITY OF INHIBITION. J Biol Chem. 2008;283:24748–59. https://doi.org/10.1074/jbc.M800352200.

96. Bochkov VN, Kadl A, Huber J, Gruber F, Binder BR, Leitinger N. Protective role of phospholipid oxidation products in endotoxin-induced tissue damage. Nature. 2002;419:77–81. https://doi.org/10.1038/nature01023.
97. Miller YI, Viriyakosol S, Binder CJ, Feramisco JR, Kirkland TN, Witztum JL. Minimally modified LDL binds to CD14, induces macrophage spreading via TLR4/MD-2, and inhibits phagocytosis of apoptotic cells. J Biol Chem. 2003;278:1561–8. https://doi.org/10.1074/jbc.M209634200.
98. Devitt A, Moffatt OD, Raykundalia C, Capra JD, Simmons DL, Gregory CD. Human CD14 mediates recognition and phagocytosis of apoptotic cells. Nature. 1998;392:505–9. https://doi.org/10.1038/33169.
99. Moffatt OD, Devitt A, Bell ED, Simmons DL, Gregory CD. Macrophage recognition of ICAM-3 on apoptotic leukocytes. J Immunol (Baltim, Md: 1950). 1999;162:6800–10.
100. Devitt A, Pierce S, Oldreive C, Shingler WH, Gregory CD. CD14-dependent clearance of apoptotic cells by human macrophages: the role of phosphatidylserine. Cell Death Differ. 2003;10:371–82. https://doi.org/10.1038/sj.cdd.4401168.
101. Devitt A, Parker KG, Ogden CA, Oldreive C, Clay MF, Melville LA, Bellamy CO, Lacy-Hulbert A, Gangloff SC, Goyert SM, Gregory CD. Persistence of apoptotic cells without autoimmune disease or inflammation in CD14−/− mice. J Cell Biol. 2004;167:1161–70. https://doi.org/10.1083/jcb.200410057.
102. Thomas L, Bielemeier A, Lambert PA, Darveau RP, Marshall LJ, Devitt A. The N-terminus of CD14 acts to bind apoptotic cells and confers rapid-tethering capabilities on non-myeloid cells. PLoS One. 2013;8:e70691. https://doi.org/10.1371/journal.pone.0070691.
103. Kol A, Lichtman AH, Finberg RW, Libby P, Kurt-Jones EA. Cutting edge: heat shock protein (HSP) 60 activates the innate immune response: CD14 is an essential receptor for HSP60 activation of mononuclear cells. J Immunol. 2000;164:13–7. https://doi.org/10.4049/jimmunol.164.1.13.
104. Asea A, Kraeft S-K, Kurt-Jones EA, Stevenson MA, Chen LB, Finberg RW, Koo GC, Calderwood SK. HSP70 stimulates cytokine production through a CD14-dependant pathway, demonstrating its dual role as a chaperone and cytokine. Nat Med. 2000;6:435–42. https://doi.org/10.1038/74697.
105. Asea A, Rehli M, Kabingu E, Boch JA, Baré O, Auron PE, Stevenson MA, Calderwood SK. Novel signal transduction pathway utilized by extracellular HSP70 ROLE OF toll-LIKE RECEPTOR (TLR) 2 AND TLR4. J Biol Chem. 2002;277:15028–34. https://doi.org/10.1074/jbc.M200497200.
106. Qin K, Ma S, Li H, Wu M, Sun Y, Fu M, Guo Z, Zhu H, Gong F, Lei P, Shen G. GRP78 impairs production of lipopolysaccharide-induced cytokines by interaction with CD14. Front Immunol. 2017;8:579. https://doi.org/10.3389/fimmu.2017.00579.
107. Zhao W, Beers DR, Bell S, Wang J, Wen S, Baloh RH, Appel SH. TDP-43 activates microglia through NF-κB and NLRP3 inflammasome. Exp Neurol. 2015;273:24–35. https://doi.org/10.1016/j.expneurol.2015.07.019.
108. Fassbender K, Walter S, Kühl S, Landmann R, Ishii K, Bertsch T, Stalder AK, Muehlhauser F, Liu Y, Ulmer AJ, Rivest S, Lentschat A, Gulbins E, Jucker M, Staufenbiel M, Brechtel K, Walter J, Multhaup G, Penke B, Adachi Y, Hartmann T, Beyreuther K. The LPS receptor (CD14) links innate immunity with Alzheimer's disease. FASEB J. 2004;18:203–5. https://doi.org/10.1096/fj.03-0364fje.
109. Liu Y, Walter S, Stagi M, Cherny D, Letiembre M, Schulz-Schaeffer W, Heine H, Penke B, Neumann H, Fassbender K. LPS receptor (CD14): a receptor for phagocytosis of Alzheimer's amyloid peptide. Brain J Neurol. 2005;128:1778–89. https://doi.org/10.1093/brain/awh531.
110. Reed-Geaghan EG, Savage JC, Hise AG, Landreth GE. CD14 and toll-like receptors 2 and 4 are required for fibrillar A{beta}-stimulated microglial activation. J Neurosci. 2009;29:11982–92. https://doi.org/10.1523/JNEUROSCI.3158-09.2009.
111. Reed-Geaghan EG, Reed QW, Cramer PE, Landreth GE. Deletion of CD14 attenuates Alzheimer's disease pathology by influencing the brain's inflammatory milieu. J Neurosci. 2010;30:15369–73. https://doi.org/10.1523/JNEUROSCI.2637-10.2010.

112. Fujikura M, Iwahara N, Hisahara S, Kawamata J, Matsumura A, Yokokawa K, Saito T, Manabe T, Matsushita T, Suzuki S, Shimohama S. CD14 and toll-like receptor 4 promote Fibrillar Aβ42 uptake by microglia through A Clathrin-Mediated pathway. J Alzheimers Dis JAD. 2019;68:323–37. https://doi.org/10.3233/JAD-180904.
113. Sano H, Sohma H, Muta T, Nomura S, Voelker DR, Kuroki Y. Pulmonary surfactant protein a modulates the cellular response to smooth and rough lipopolysaccharides by interaction with CD14. J Immunol (Baltim, Md: 1950). 1999;163:387–95.
114. Sano H, Chiba H, Iwaki D, Sohma H, Voelker DR, Kuroki Y. Surfactant proteins A and D bind CD14 by different mechanisms. J Biol Chem. 2000;275:22442–51. https://doi.org/10.1074/jbc.M001107200.
115. Augusto LA, Synguelakis M, Johansson J, Pedron T, Girard R, Chaby R. Interaction of pulmonary surfactant protein C with CD14 and lipopolysaccharide. Infect Immun. 2003;71:61–7. https://doi.org/10.1128/IAI.71.1.61-67.2003.
116. Yang L, Carrillo M, Wu YM, DiAngelo SL, Silveyra P, Umstead TM, Halstead ES, Davies ML, Hu S, Floros J, McCormack FX, Christensen ND, Chroneos ZC. SP-R210 (Myo18A) isoforms as intrinsic modulators of macrophage priming and activation. PLoS One. 2015;10:e0126576. https://doi.org/10.1371/journal.pone.0126576.
117. Baveye S, Elass E, Fernig DG, Blanquart C, Mazurier J, Legrand D. Human lactoferrin interacts with soluble CD14 and inhibits expression of endothelial adhesion molecules, E-selectin and ICAM-1, induced by the CD14-lipopolysaccharide complex. Infect Immun. 2000;68:6519–25. https://doi.org/10.1128/iai.68.12.6519-6525.2000.
118. Saenger T, Vordenbäumen S, Genich S, Haidar S, Schulte M, Nienberg C, Bleck E, Schneider M, Jose J. Human αS1-casein induces IL-8 secretion by binding to the ecto-domain of the TLR4/MD2 receptor complex. Biochim Biophys Acta Gen Subj. 2019;1863:632–43. https://doi.org/10.1016/j.bbagen.2018.12.007.
119. Kim S, Takahashi H, Lin W-W, Descargues P, Grivennikov S, Kim Y, Luo J-L, Karin M. Carcinoma-produced factors activate myeloid cells through TLR2 to stimulate metastasis. Nature. 2009;457:102–6. https://doi.org/10.1038/nature07623.
120. Schaefer L, Babelova A, Kiss E, Hausser H-J, Baliova M, Krzyzankova M, Marsche G, Young MF, Mihalik D, Götte M, Malle E, Schaefer RM, Gröne H-J. The matrix component biglycan is proinflammatory and signals through toll-like receptors 4 and 2 in macrophages. J Clin Invest. 2005;115:2223–33. https://doi.org/10.1172/JCI23755.
121. Roedig H, Nastase MV, Frey H, Moreth K, Zeng-Brouwers J, Poluzzi C, Hsieh LT-H, Brandts C, Fulda S, Wygrecka M, Schaefer L. Biglycan is a new high-affinity ligand for CD14 in macrophages. Matrix Biol. 2019;77:4–22. https://doi.org/10.1016/j.matbio.2018.05.006.
122. He Z, Riva M, Björk P, Swärd K, Mörgelin M, Leanderson T, Ivars F. CD14 is a co-receptor for TLR4 in the S100A9-induced pro-inflammatory response in monocytes. PLoS One. 2016;11:e0156377. https://doi.org/10.1371/journal.pone.0156377.
123. Cavaillon J-M, Marie C, Caroff M, Ledur A, Godard I, Poulain D, Fitting C, Haeffner-Cavaillon N. CD14/LPS receptor exhibits lectin-like properties. J Endotoxin Res. 1996;3:471–80. https://doi.org/10.1177/096805199600300605.
124. Wang J, Huo K, Ma L, Tang L, Li D, Huang X, Yuan Y, Li C, Wang W, Guan W, Chen H, Jin C, Wei J, Zhang W, Yang Y, Liu Q, Zhou Y, Zhang C, Wu Z, Xu W, Zhang Y, Liu T, Yu D, Zhang Y, Chen L, Zhu D, Zhong X, Kang L, Gan X, Yu X, Ma Q, Yan J, Zhou L, Liu Z, Zhu Y, Zhou T, He F, Yang X. Toward an understanding of the protein interaction network of the human liver. Mol Syst Biol. 2011;7:536. https://doi.org/10.1038/msb.2011.67.
125. Hein MY, Hubner NC, Poser I, Cox J, Nagaraj N, Toyoda Y, Gak IA, Weisswange I, Mansfeld J, Buchholz F, Hyman AA, Mann M. A human Interactome in three quantitative dimensions organized by Stoichiometries and abundances. Cell. 2015;163:712–23. https://doi.org/10.1016/j.cell.2015.09.053.
126. Baumann CL, Aspalter IM, Sharif O, Pichlmair A, Blüml S, Grebien F, Bruckner M, Pasierbek P, Aumayr K, Planyavsky M, Bennett KL, Colinge J, Knapp S, Superti-Furga

G. CD14 is a coreceptor of Toll-like receptors 7 and 9. J Exp Med. 2010;207:2689–701. https://doi.org/10.1084/jem.20101111.

127. Li J, Ahmet F, O'Keeffe M, Lahoud MH, Heath WR, Caminschi I. CD14 is not involved in the uptake of synthetic CpG oligonucleotides. Mol Immunol. 2017;81:52–8. https://doi.org/10.1016/j.molimm.2016.11.015.
128. Bae YS, Lee JH, Choi SH, Kim S, Almazan F, Witztum JL, Miller YI. Macrophages generate reactive oxygen species in response to minimally oxidized low-density lipoprotein: toll-like receptor 4- and spleen tyrosine kinase-dependent activation of NADPH oxidase 2. Circ Res. 2009;104:210–8, 21p following 218. https://doi.org/10.1161/CIRCRESAHA.108.181040.
129. Choi S-H, Wiesner P, Almazan F, Kim J, Miller YI. Spleen tyrosine kinase regulates AP-1 dependent transcriptional response to minimally oxidized LDL. PLoS One. 2012;7:e32378. https://doi.org/10.1371/journal.pone.0032378.
130. Choi S-H, Harkewicz R, Lee JH, Boullier A, Almazan F, Li AC, Witztum JL, Bae YS, Miller YI. Lipoprotein accumulation in macrophages via toll-like receptor-4-dependent fluid phase uptake. Circ Res. 2009;104:1355–63. https://doi.org/10.1161/CIRCRESAHA.108.192880.
131. Kojima Y, Weissman IL, Leeper NJ, Leeper NJ. The role of Efferocytosis in atherosclerosis. Circulation. 2017;135:476–89. https://doi.org/10.1161/CIRCULATIONAHA.116.025684.
132. Zanoni I, Tan Y, Di Gioia M, Broggi A, Ruan J, Shi J, Donado CA, Shao F, Wu H, Springstead JR, Kagan JC. An endogenous caspase-11 ligand elicits interleukin-1 release from living dendritic cells. Science. 2016;352:1232–6. https://doi.org/10.1126/science.aaf3036.
133. Di Gioia M, Spreafico R, Springstead JR, Mendelson MM, Joehanes R, Levy D, Zanoni I. Endogenous oxidized phospholipids reprogram cellular metabolism and boost hyperinflammation. Nat Immunol. 2020;21:42–53. https://doi.org/10.1038/s41590-019-0539-2.
134. Geng S, Chen K, Yuan R, Peng L, Maitra U, Diao N, Chen C, Zhang Y, Hu Y, Qi C-F, Pierce S, Ling W, Xiong H, Li L. The persistence of low-grade inflammatory monocytes contributes to aggravated atherosclerosis. Nat Commun. 2016;7:13436. https://doi.org/10.1038/ncomms13436.
135. Quintana FJ, Cohen IR. Heat shock proteins as endogenous adjuvants in sterile and septic inflammation. J Immunol (Baltim, Md: 1950). 2005;175:2777–82. https://doi.org/10.4049/jimmunol.175.5.2777.
136. Mann JR, Gleixner AM, Mauna JC, Gomes E, DeChellis-Marks MR, Needham PG, Copley KE, Hurtle B, Portz B, Pyles NJ, Guo L, Calder CB, Wills ZP, Pandey UB, Kofler JK, Brodsky JL, Thathiah A, Shorter J, Donnelly CJ. RNA binding antagonizes neurotoxic phase transitions of TDP-43. Neuron. 2019;102:321–338.e8. https://doi.org/10.1016/j.neuron.2019.01.048.
137. Neumann M, Sampathu DM, Kwong LK, Truax AC, Micsenyi MC, Chou TT, Bruce J, Schuck T, Grossman M, Clark CM, McCluskey LF, Miller BL, Masliah E, Mackenzie IR, Feldman H, Feiden W, Kretzschmar HA, Trojanowski JQ, Lee VM-Y. Ubiquitinated TDP-43 in frontotemporal lobar degeneration and amyotrophic lateral sclerosis. Science. 2006;314:130–3. https://doi.org/10.1126/science.1134108.
138. Uryu K, Nakashima-Yasuda H, Forman MS, Kwong LK, Clark CM, Grossman M, Miller BL, Kretzschmar HA, Lee VM-Y, Trojanowski JQ, Neumann M. Concomitant TAR-DNA-binding protein 43 pathology is present in Alzheimer disease and corticobasal degeneration but not in other tauopathies. J Neuropathol Exp Neurol. 2008;67:555–64. https://doi.org/10.1097/NEN.0b013e31817713b5.
139. McKee AC, Gavett BE, Stern RA, Nowinski CJ, Cantu RC, Kowall NW, Perl DP, Hedley-Whyte ET, Price B, Sullivan C, Morin P, Lee H-S, Kubilus CA, Daneshvar DH, Wulff M, Budson AE. TDP-43 Proteinopathy and motor neuron disease in chronic traumatic encephalopathy. J Neuropathol Exp Neurol. 2010;69:918–29. https://doi.org/10.1097/NEN.0b013e3181ee7d85.

Toll-*Like* Receptor 4 Interactions with *Neisseria*

Myron Christodoulides

Abstract *Neisseria meningitidis* (meningococcus) causes meningitis and sepsis and *Neisseria gonorrhoeae* (gonococcus) causes the sexually transmitted disease gonorrhoea. Both pathogens contain lipooligosaccharide (LOS) in their outer membrane (OM). LOS interacts as a monomer with the MD-2/*Toll*-like receptor (TLR4) complex found in eukaryotic cell membranes. Binding recognition triggers signalling pathways that stimulate host innate and inflammatory responses. In this review, we examine the molecular mechanism of *Neisseria* LOS interaction with TLR4; the role of TLR4 in meningococcal and gonococcal infections; whether LOS-antagonists can be inhibition therapies; and whether *Neisseria* vaccine-induced immune responses can exploit TLR4.

Keywords Neisseria · Meningitis · Sepsis · Gonorrhoea · Lipooligosaccharide · Polymorphisms · Inhibition therapy · Vaccine

1 Introduction

The Gram-negative bacterium *Neisseria meningitidis* (meningococcus) causes systemic meningococcal disease (SMD) and the sister organism *Neisseria gonorrhoeae* (gonococcus) causes the sexually transmitted disease gonorrhoea. The characteristics of SMD are compartmentalized bacterial growth and inflammation in the blood (sepsis) and cerebrospinal fluid (CSF meningitis) [1]. Gonorrhoea is an inflammatory response in the genital tract: in men, infection of the urethra causes urethritis and painful discharge, and in women, localised infection of the ectocervix and endocervix leads to a mucopurulent cervicitis. However, in approximately 10–25%

M. Christodoulides (✉)
Neisseria Research Group, Molecular Microbiology, School of Clinical and Experimental Sciences, Sir Henry Wellcome Laboratories, University of Southampton Faculty of Medicine, Southampton, UK
e-mail: mc4@soton.ac.uk

C. Rossetti, F. Peri (eds.), *The Role of Toll-Like Receptor 4 in Infectious and Non Infectious Inflammation*, Progress in Inflammation Research 87,
https://doi.org/10.1007/978-3-030-56319-6_5

of untreated women, gonococci can ascend into the upper reproductive tract and initiate pelvic inflammatory disease (PID) [2]. Gonococci can also cause anorectal and pharyngeal infections, more rarely disseminated infection, and *ophthalmia neonatorum* (conjunctivitis) in neonates. Recognition of *Neisseria*-associated molecular patterns (NAMPs) by *Toll*-like receptors (TLRs), including TLR4, is critical for host innate and inflammatory defence against these organisms.

2 Mechanism of TLR4 Interaction with *Neisseria* LOS

The outside layer of the outer membrane (OM) of the *Neisseria* is composed of lipooligosaccharide (LOS), proteins and phospholipids (principally phosphatidylethanolamine and varying amounts of phosphatidylglycerol, cardiolipin and phosphatidate) [3], whereas the inner layer is composed of phospholipids and regulatory proteins for nutrients and metabolic products. LOS is composed of a lipid A containing hydroxy fatty-acid chains and phosphoethanolamine (PEA), a core oligosaccharide containing two 3-deoxy-D-manno-oct-2-ulosonic acids (KDO) and two heptose residues, and highly variable short oligosaccharides. LOS is structurally distinct from the lipopolysaccharide (LPS) of Gram-negative enteric bacilli because it lacks a repeating polysaccharide O-side chain [4]. Gonococci and meningococci possess the *lgt* genes encoding LOS with high genetic diversity [5] and LOS phase and antigenic variations lead to different chain oligosaccharide and inner core structures and a classification system of LOS immunotypes (L1-L12 in meningococci). The terminal lacto-*N*-neotetraose at the non-reducing end of the LOS oligosaccharide is similar to the mammalian glycosphingolipid paragloboside and this molecular mimicry is important for evading human immune responses. The lacto-*N*-neotetraose also can be modified by sialylation using endogenous sialyltransferase enzyme and an appropriate exogenous sialic acid substrate in vivo, or in vitro, for example, with cytidine 5′-monophospho-*N*-acetylneuraminic acid. Sialylation of LOS enables *Neisseriae* to resist human serum complement and inhibits both classical and alternative complement pathways [6].

Neisseria LOS interacts sequentially with LPS-binding protein (LBP), CD14, MD-2 and TLR4 (Fig. 1). The LBP-LOS complex combines with CD14 on the cell surface and monomeric LOS is transferred. LBP-CD14 can extract OM phospholipids and LOS, but only LOS-CD14 reacts with MD-2 and activates the cells via TLR4 [7]. The LBP-CD14-dependent extraction and transfer of LOS monomer is accompanied by increased exposure of lipid A fatty acyl chains, which are then sequestered when LOS binds to MD-2 [8]. Direct interaction between lipid A in monomeric LOS with MD-2 activates TLR4, which is followed by TLR4 dimerization, a requirement for recruiting signal transduction pathway molecules. Monomeric MD-2 is the active species for reactivity between CD14 and TLR4 for LOS-induced activation of TLR4 [9]. There is a correlation between variations in the structure of lipid A, which can influence the binding affinity of meningococcal LOS for MD-2, and subsequent TLR4 activation that leads to cytokine production by human

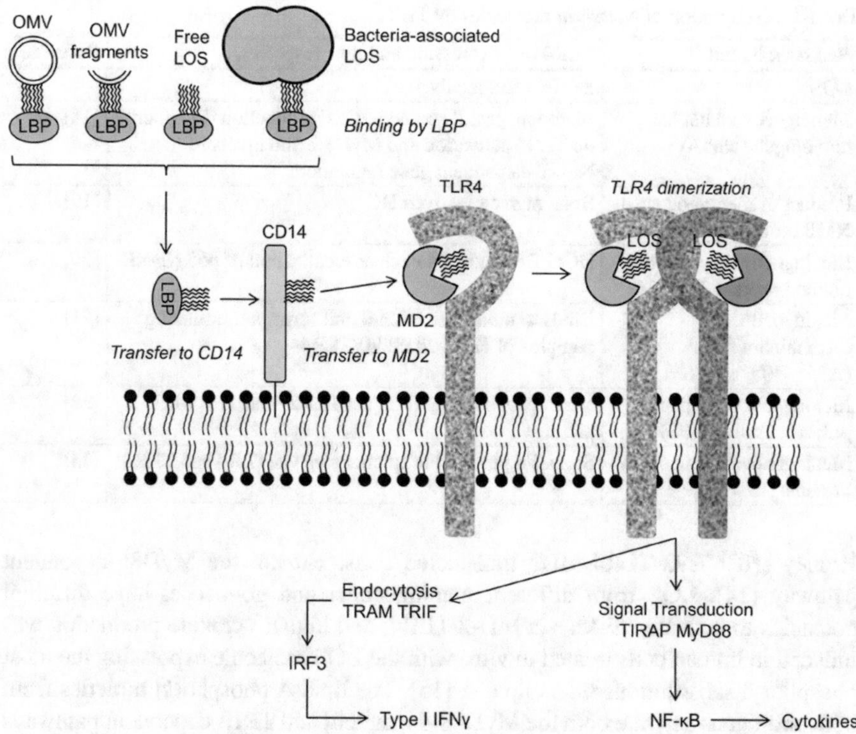

Fig. 1 Cartoon of *Neisseria* LOS interactions with TLR4. LOS is present in the OM, in OM blebs, OM fragments and as free LOS. LPS-binding protein (LBP) binds free LOS and fragmented OM-LOS preferentially and transfers LOS to membrane-anchored CD14. CD14 transfers single molecules of LOS to MD-2, which then interacts with membrane-bound TLR4 and induces TLR4 dimerization. This process of dimerization is necessary for signal transduction from the membrane via TIRAP/MyD88, which leads to NF-κB activation and cytokine production. In addition, endocytosis of the LOS-MD-2/TLR4 complex triggers TRAM/TRIF signalling pathways to IRF3 and production of Type 1 IFNγ

macrophages [10, 11]. The KDO2 moiety linked to meningococcal lipid A is identified as the structural unit required for maximal activation of the macrophage TLR4 pathway [10]. Fine structural differences in LOS can influence signalling through MD-2/TLR4, for example, tetra-, penta- and hexa-acylated LOS structures can form LOS-MD-2 complexes, but the hexa-acylated forms are the most potent TLR4 agonists [11]. NMR studies show that arrangement of amino acid Phe(126) at the mouth of the hydrophobic cavity of MD-2 can act as a 'hydrophobic switch', which drives LOS-dependent fatty acyl contacts that are required for TLR4 dimerization and activation [12].

After binding to TLR4, meningococcal LOS can induce both MyD88-dependent and MyD88-independent (TRIF-dependent) signalling pathways in human cells [13]. Interestingly, diverse meningococcal strains show differences in their ability to activate the TLR4-induced MyD88-independent pathway in Human Embryonic

Table 1 Recognition of *Neisseria* molecules by TLR4

Neisseria ligand	TLR4 cell expression and host response	References
LOS	See text for details	
Meningococcal hia/hsf homologue (NhhA)	Macrophages; IL-6 and C-CSF production dependent on TLR4 activation and MyD88, and involved NF-κB-dependent gene regulation	[18]
Product of meningococcal NMB1468 (Ag-473)	Bone marrow-derived DC	[19]
Meningococcal penicillin-binding protein 2 (PBP2)	DC; PBP2 induced nuclear localization of p65 NF-κB	[20]
Meningococcal recombinant NadA (Δ351–405)	Binds to monocyte HSP90 and forms a transducing complex of HSP90/HSP70/TLR4	[21]
Meningococcal capsule polysaccharide (CPS)	Macrophage recognition via TLR2- and TLR4-MD-2 pathways	[22]
Meningococcal heat shock proteins (HSPs)	Signalling in murine splenocytes via TLR4 and TLR2	[23]

Kidney (HEK)293-TLR4-MD2 transfected cells, but not the MyD88-dependent pathway [14]. LOS from different meningococci and gonococci have different potencies to activate NF-kB via MD-2/TLR4, and higher cytokine production was induced in human cells treated in vitro with the LOS molecule expressing the most phosphoryl substitutions on its lipid A [15]. The lipid A phosphoryl moieties from both pathogens activate both the MyD88-dependent and TRIF-dependent pathways through NF-kB and IRF3 transcription factors, respectively (Fig. 1) [15]. *Neisseria* LOS-induced signalling probably occurs via MyD88 from the plasma membrane and/or the TRIF-dependent pathway via endosomal compartmentalization of the LOS-TLR4-MD2 complex [16]. Human dendritic cells (DC) contained TLR4 in close association with the Golgi complex, and co-localisation with α-tubulin microtubules suggested that microtubules act as transport pathways for TLR vesicles [17].

Activation of TLR4 by non-LOS Pathogen Associated Molecular Patterns (PAMPs) has also been reported (Table 1), although the possible presence of minor contaminating LOS/LPS in the preparations must temper these interpretations. It is worth noting, nevertheless, the hypothesis that a NadA transducing signal was necessary for the immuno-stimulating activity of NadA present in the Bexsero meningitis vaccine.

3 TLR4 in *Neisseria meningitidis* Infection

LOS is the main toxic component that triggers sepsis and meningitis and recognition of LOS by the TLR4 complex leads to compartmentalised pro-inflammatory cytokine/chemokine responses [24]. During sepsis, CD14 is required for low to medium concentrations of LOS to activate the TLR4/MD-2 complex, whereas the complex can be activated directly without CD14 by high LOS concentrations [25]. Meningococci activate vessel endothelial cells, monocytes and polymorphonuclear leucocytes (PMNL) in the blood, and activate the leptomeninges, vessel endothelial cells, sentinel macrophages and infiltrating PMNLs in the CSF, leading to the production of TNFα, IL-1β, IL-6 and IL-8 in both compartments [1, 26]. Experiments using antibody-mediated blockage of the CD14-TLR4 receptor complex [27–31] and siRNA-mediated TLR4 silencing [32] demonstrate the essential role of TLR4 in LOS recognition.

LPS activation of murine TLR4 induces up-regulation of Class A SR MARCO (macrophage receptor with a collagenous structure) and SR-A (scavenger receptor A), which may regulate TLR-mediated immune responses and be important during the early stages of infection. However, these receptors are not important for survival of mice infected with meningococci but do contribute to meningococcal capture and clearance [33]. The pro-inflammatory phase in SMD is followed by an anti-inflammatory phase, and the characteristic presence of IL-10 in sepsis patients increases the expression of *tlr4*-associated genes, including those for *md-2* and soluble (s)*cd14* [34]. Over-expression of sCD14, in particular, could reduce LOS lethality by sequestering LOS in the circulation and preventing TLR4/MD-2 interactions [34]. Mammals can potentially also regulate innate immune responses to bacterial infection by chemical modification of LOS, for example, through the host enzyme acyloxyacyl hydrolase, which deacylates LOS [35], or via CD200 ligand induction by the LOS-TLR4-mediated MyD88 signalling pathway [36]. Conversely, some meningococcal disease isolates have natural inactivating mutations in the *lpxL1* gene, and lipid A mutants show reduced virulence in humans [37]. The mutations generate lipid A structures that are less efficient in activating TLR4 and provide the organisms with a mechanism to escape innate immune system recognition [37].

3.1 *Influence of TLR4 Functional Polymorphisms on SMD*

Human gene polymorphisms that affect TLR4 expression and function might influence susceptibility to Gram-negative bacterial infection and the severity of diseases such as sepsis. The earliest investigation of this hypothesis showed that the single nucleotide polymorphism (SNP) Asp299Gly in the human *tlr4* gene that caused reduced expression and function of TLR4 did not influence susceptibility to, or severity of, SMD [38]. A study of serogroup A meningitis in Gambian children also reported no association between the TLR4 SNPs Asp299Gly (rs4986790) and Thr399Ile (rs4986791) and meningitis [39]. There was no association between both these TLR4 SNPs and susceptibility to meningitis in a small study of children with meningococcal meningitis and their uninfected family members [40]. In addition, there was no significant influence of both TLR4 Asp299Gly and CD14 C-159T SNPs on the risk of developing invasive meningococcal disease in a cohort of surviving patients [41]. More recently, a prospective descriptive study of TLR2/TLR4/CD14 polymorphisms and predisposition to infection by meningococci and pneumococci found that the TLR2 p.753Q allele and the CD14 c.-159T (C > T) allele, but not the TLR4 p.D299G allele, were more frequent in children with invasive disease than in the control group [42]. In vitro, monocytes from individuals heterozygous for the TLR4 gene mutations Asp299Gly and Thr399Ile also showed no deficit in recognising LPS/LOS [43]. However, there was evidence for a protective role in invasive bacterial disease, including meningococcal infections, for TLR4 rs2149356 G/T [44]. Overall, these studies suggest that rare TLR4 variants are not major determinants of meningococcal disease, although larger cohort studies may be warranted.

3.2 TLR4-LOS Inhibition Therapies

The literature on targeting TLR4 with LOS-antagonists as adjunctive treatments for *Neisseria*-induced inflammatory responses is sparse (Table 2). Only the LPS-antagonists Eritoran (E5554) and bactericidal/permeability increasing (BPI) protein have entered large clinical trials and neither showed any significant impact on sepsis mortality rates [46, 47] (Table 2). However, it is important to stress that during SMD, innate recognition of non-LOS modulins independent of TLR4/MD-2 interactions and signalling also drive inflammatory responses. This is amply demonstrated in several studies in vitro using, for example, human and mouse monocytes/macrophages, granulocytes and DC [51–53] and human leptomeningeal cells [54]. Thus, considered attempts to inhibit TLR4 activation alone to reduce inflammation may be of limited value.

4 TLR4 in *Neisseria gonorrhoeae* Infection

There are cursory studies on the role of TLR4 in gonorrhoea. Mucosal epithelial cells isolated from the normal human vagina, ectocervix and endocervix do not express TLR4 and MD-2 mRNA, which explains their unresponsiveness to protein-free LOS preparations from gonococci [55, 56]. However, lower reproductive tract epithelial cells do respond to whole bacteria and bacterial lysates, demonstrating that other PAMPS and non-LOS pathogen recognition receptors (PRRs) are involved in the host innate immune response [57]. In addition, TLR4 expression on residential and infiltrating immune cells found in whole organism models probably contributes to recognition of gonococcal LOS during infection.

Several lines of evidence suggest a protective role for TLR4 during gonococcal infection. TLR4 activation may be indirectly involved in gonococcal adherence to host cells: LOS-containing OM blebs induced TLR4-dependent activation of NF-κB in cultured human umbilical vein endothelial cells, which in turn upregulated the expression of the CEACAM1 receptor to enable Opa-mediated gonococcal adherence [58]. Activation of CEACAM3 receptor on neutrophils by Opa binding leads to gonococcal engulfment as well as cytokine production in cooperation with other

Table 2 TLR4-LOS inhibition therapies

Drug	Mechanism of action	References
Steroid dexamethasone	Inhibits meningococcal LOS-induced TLR signalling by interfering with MyD88-dependent signal transduction	[45]
Eritoran (E5554)	Synthetic lipid A antagonist that blocks LPS binding at MD2-TLR4	[46]
Bactericidal/permeability increasing (BPI) protein	Inhibits transfer of LPS to CD14 and hence TLR4-mediated cell activation. Produces aggregates of meningococcal LOS-BPI that cells can take up independently of the CD14 and TLR4 machinery, without inducing inflammatory responses and directing the LOS for clearance	[47]
Meningococcal lipid A mutant LOS molecules lpxL1 (penta-acylated) and lpxL2 (tetra-acylated)	Inhibits TLR4-dependent cytokine production	[48]
Cyp, a selective TLR4-MD-2 antagonist derived from the cyanobacterium *Oscillatoria planktothrix* FP1	Competitively inhibits LOS interactions with TLR4 and reduces NF-κB activation. Inhibits cytokine production from a human whole blood model treated with pure meningococcal LOS and OMV, and infected with live bacteria	[49]
Rhodobacter sphaeroides diphosphoryl lipid A (RsDPLA)	Blocks TLR4	[27]
Synthetic amphiphilic glycolipids containing a positively charged amino group or an ammonium salt and two lipophilic chains	Inhibit LPS-induced TLR4 activation on TLR4-transfected HEK cells in vitro and LPS-induced septic shock in mice. These compounds inhibit TLR4 activation by meningococcal LOS by competitively occupying CD14 and reducing the level of delivery of activating LOS to the MD2-TLR4 complex	[50]

PRRs. Notably, CEACAM3-dependent signals integrated with the PRR network via its engagement of Malt1 and Bcl10 molecules. In particular, Bcl10 acts as a mediator and contributes to the synergistic amplification of CEACAM3 and TLR4 signals, leading to cytokine production [59]. Regulating the magnitude of the innate immune response during gonococcal infection to avoid excessive inflammation is also important and probably involves the contribution of miRNA expression. Activation of TLR4 by LOS induced the expression of the anti-inflammatory microRNA miR-178 in macrophages in vitro [60]. Decreased miRNA-178 expression was associated with increased gonococcal burden during infection, and miRNA718 controls both TLR4 and inflammatory cytokine signalling via a negative-feedback regulation loop that down-regulates TLR4, IRAK1 and NF-κB.

Further evidence for a protective role for TLR4 in gonorrhoea comes from one study in which female mice carrying the *Lps(d)* mutation in TLR4, which renders them unresponsive to endotoxin, were inoculated intravaginally with *N. gonorrhoeae* [61]. There was no difference in the duration of gonococcal colonization, but

Lps(d) mice had a significantly higher burden of gonococci, which was coincident with a large PMNL influx and upregulation of TNFα, IL-1β, MIP-2 and KC inflammatory cytokines. Infected *Lps(d)* mice showed a decrease in IL-17, which suggests that Th17 responses may be more dependent on TLR4 signalling in vivo. Furthermore, *Lps(d)* mice showed defective PMNL-mediated and complement-independent serum killing of gonococci [61].

Gonococcal LOS modification(s) influences TLR4 activation. PEA decoration of LOS lipid A is a mechanism whereby gonococci increase their resistance to complement-mediated bacteriolysis and the effects of cationic antimicrobial peptides (CAMPs). PEA modification enhanced both MyD88- and TRIF-dependent TLR4 signalling pathways in vitro [15] and correspondingly, pure LOS with PEA-deficient lipid A induced lower levels of NF-κB in human TLR4-HEK cells than wild-type LOS [62]. Host-derived CAMPS such as cathelicidin also reduced further the capacity of PEA-deficient LOS to interact with TLR4 [62]. An aspect of *Neisseria* LOS interactions with human cells is the phenomenon of tolerance. Treatment of monocyte cells with pathogenic *Neisseria* and pure LOS, after a previous exposure, induces tolerance, manifested by reduced TNFα and IL-1β cytokine expression [63]. Significantly, the modified LOS molecules that varied in their potential to activate TLR4 [15] also had variable ability to induce tolerance, and microRNA-146a regulated this response [63].

There is compelling clinical and epidemiological evidence of a positive correlation between *N. gonorrhoeae* infection and HIV infection and transmission. However, TLR4 agonists such as gonococci did not enhance HIV-1 infection of primary resting CD4(+) T cells after viral entry, with enhancement of viral infection dependent instead on TLR2 activation [64]. Gonococci release heptose monophosphate that activates CD4+ T lymphocytes and induces a NF-κB-dependent transcriptional response that drives HIV-1 expression and virus production. However, this is not via TLR4, as mutation in the gonococcal ADP-heptose biosynthesis gene *hldA*, which produces a truncated LOS that is still bioactive in a TLR4 reporter-based assay, generated a gonococcal mutant that did not induce HIV-1 expression in CD4+ T lymphocytes [65]. Co-infection of plasmacytoid DC with HIV-1 and gonococci inhibited viral replication, but via TLR9-dependent Interferon alpha production [66]. However, recent studies show that complete IFN induction by gonococcal infection of human and murine cells in vitro depends on both TLR4 and the binding of double-stranded bacterial DNA with cytosolic enzyme cyclic-guanosine monophosphate–adenosine monophosphate (GMP-AMP) synthase (cGAS). This produces 2′3’-cGAMP and triggers STING/TBK-1/IRF3 activation, resulting in type I IFN expression [67].

5 Exploiting TLR4 in *Neisseria* Vaccine-Induced Immune Responses

OM Vesicle (OMV) vaccines have successfully controlled clonal outbreaks of serogroup B meningococcal disease [68] and their preparation involves treatment with the detergent sodium deoxycholate (NaDOC) to reduce LOS content and vaccine reactogenicity. TLR4 activation contributes to OMV vaccine immunogenicity [69] probably via recognition of residual LOS. However, NaDOC extraction impacts the OMV proteome, so alternative genetic modification strategies have been developed to detoxify LOS, especially the lipid A moiety, or eliminate it entirely. For example, inactivating the *lpxA* gene produces LOS-deficient meningococci [70], but this impacts the immunogenicity of major OM proteins, due to the absence of inherent LOS adjuvanticity. Mutations, for example, in the *lpxL1* gene, produce penta-acylated lipid A that has reduced endotoxicity, compared with the wild-type hexa-acylated lipid A, but this strategy also results in a loss of inherent adjuvanticity, as the penta-acylated variety is a poor activator of human MD-2/TLR4 [71, 72]. Unglycosylated meningococcal lipid A may be a candidate adjuvant since it is a weak agonist for MD-2/TLR4 in human macrophages [73]. LOS endotoxicity can be modulated by combinatorial bio-engineering to produce LOS-derivatives with broad ranges of TLR4 activation that can be exploited as vaccine adjuvants [74]. In a different strategy, the addition of TLR4 agonists such as non-toxic monophosphoryl lipid A and a mutant penta-acylated *lpxL1* LOS increased total IgG antibody against LOS-deficient OMV and induced higher serum bactericidal titres compared to the LOS-depleted OMV alone [75]. Conversely, formulation in liposomes containing recombinant transferrin-binding protein B reduces the reactogenicity of native meningococcal LOS and increases LOS immunogenicity [76].

6 Conclusions

Inflammatory responses to *Neisseria* infections are complex and TLR4 is essential for innate immune recognition of LOS. TLR4 plays a role for early containment of infection, although regulation of the TLR4 receptor complex, its signalling pathways and host defence is unclear. Understanding how to regulate NAMP recognition that promotes *Neisseria* clearance, without inducing overwhelming inflammatory responses that are detrimental to the host, could lead to the development of new adjunctive therapies for use during infections.

References

1. Brandtzaeg P, van Deuren M. Classification and pathogenesis of meningococcal infections. In: Christodoulides M, editor. *Neisseria meningitidis*: advanced methods and protocols, Methods in molecular biology, vol. 799. New York: Humana Press; 2012. p. 21–35.
2. Mitchell C, Prabhu M. Pelvic inflammatory disease: current concepts in pathogenesis, diagnosis and treatment. Infect Dis Clin N Am. 2013;27(4):793–809.
3. Rahman MM, Kolli VS, Kahler CM, et al. The membrane phospholipids of *Neisseria meningitidis* and *Neisseria gonorrhoeae* as characterized by fast atom bombardment mass spectrometry. Microbiology. 2000;146(Pt 8):1901–11.
4. Kahler CM, Stephens DS. Genetic basis for biosynthesis, structure, and function of meningococcal lipooligosaccharide (Endotoxin). Crit Rev Microbiol. 1998;24(4):281–334.
5. Zhu P, Klutch MJ, Bash MC, et al. Genetic diversity of three lgt loci for biosynthesis of lipooligosaccharide (LOS) in *Neisseria* species. Microbiology. 2002;148(Pt 6):1833–44.
6. Smith H, Parsons NJ, Cole JA. Sialylation of neisserial lipopolysaccharide: a major influence on pathogenicity. Microb Pathog. 1995;19(6):365–77.
7. Post DM, Zhang D, Eastvold JS, et al. Biochemical and functional characterization of membrane blebs purified from *Neisseria meningitidis* serogroup B. J Biol Chem. 2005;280(46):38383–94.
8. Gioannini TL, Teghanemt A, Zhang D, et al. Endotoxin-binding proteins modulate the susceptibility of bacterial endotoxin to deacylation by acyloxyacyl hydrolase. J Biol Chem. 2007;282(11):7877–84.
9. Teghanemt A, Widstrom RL, Gioannini TL, et al. Isolation of monomeric and dimeric secreted MD-2. Endotoxin.sCD14 and Toll-like receptor 4 ectodomain selectively react with the monomeric form of secreted MD-2. J Biol Chem. 2008;283(32):21881–9.
10. Zughaier SM, Tzeng YL, Zimmer SM, et al. *Neisseria meningitidis* lipooligosaccharide structure-dependent activation of the macrophage CD14/toll-like receptor 4 pathway. Infect Immun. 2004;72(1):371–80.
11. Zimmer SM, Zughaier SM, Tzeng YL, et al. Human MD-2 discrimination of meningococcal lipid A structures and activation of TLR4. Glycobiology. 2007;17(8):847–56.
12. Yu L, Phillips RL, Zhang D, et al. NMR studies of hexaacylated endotoxin bound to wild-type and F126A mutant MD-2 and MD-2.TLR4 ectodomain complexes. J Biol Chem. 2012;287(20):16346–55.
13. Zughaier SM, Zimmer SM, Datta A, et al. Differential induction of the toll-like receptor 4-MyD88-dependent and -independent signaling pathways by endotoxins. Infect Immun. 2005;73(5):2940–50.
14. Mogensen TH, Paludan SR, Kilian M, et al. Two *Neisseria meningitidis* strains with different ability to stimulate toll-like receptor 4 through the MyD88-independent pathway. Scand J Immunol. 2006;64(6):646–54.
15. Liu M, John CM, Jarvis GA. Phosphoryl moieties of lipid A from *Neisseria meningitidis* and *N. gonorrhoeae* lipooligosaccharides play an important role in activation of both MyD88- and TRIF-dependent TLR4-MD-2 signaling pathways. J Immunol. 2010;185(11):6974–84.
16. Molteni M, Gemma S, Rossetti C. The role of Toll-Like receptor 4 in infectious and noninfectious inflammation. Mediat Inflamm. 2016;2016:6978936.
17. Uronen-Hansson H, Allen J, Osman M, et al. Toll-like receptor 2 (TLR2) and TLR4 are present inside human dendritic cells, associated with microtubules and the Golgi apparatus but are not detectable on the cell surface: integrity of microtubules is required for interleukin-12 production in response to internalized bacteria. Immunology. 2004;111(2):173–8.
18. Sjolinder M, Altenbacher G, Wang X, et al. The meningococcal Adhesin NhhA provokes proinflammatory responses in macrophages via toll-like receptor 4-dependent and -independent pathways. Infect Immun. 2012;80(11):4027–33.
19. Chu CL, Yu YL, Kung YC, et al. The immunomodulatory activity of meningococcal lipoprotein Ag473 depends on the conformation made up of the lipid and protein moieties. PLoS One. 2012;7(7):e40873.

20. Hill M, Deghmane AE, Segovia M, et al. Penicillin binding proteins as danger signals: meningococcal penicillin binding protein 2 activates dendritic cells through toll-like receptor 4. PLoS One. 2011;6(10):e23995.
21. Cecchini P, Tavano R, Polverino de Laureto P, et al. The soluble recombinant *Neisseria meningitidis* adhesin NadA(Delta351-405) stimulates human monocytes by binding to extracellular Hsp90. PLoS One. 2011;6(9):e25089.
22. Zughaier SM. *Neisseria meningitidis* capsular polysaccharides induce inflammatory responses via TLR2 and TLR4-MD-2. J Leukoc Biol. 2011;89(3):469–80.
23. Han JX, Ng GZ, Cecchini P, et al. Heat shock protein complex vaccines induce antibodies against *Neisseria meningitidis* via a MyD88-independent mechanism. Vaccine. 2016;34(14):1704–11.
24. Koedel U. Toll-like receptors in bacterial meningitis. Curr Top Microbiol Immunol. 2009;336:15–40.
25. Hellerud BC, Stenvik J, Espevik T, et al. Stages of meningococcal sepsis simulated in vitro, with emphasis on complement and Toll-like receptor activation. Infect Immun. 2008;76(9):4183–9.
26. Christodoulides M. Inflammation in the subarachnoid space. In: Christodoulides M, editor. Meningitis: cellular and molecular basis. Wallingford: CABI; 2013.
27. Bjerre A, Brusletto B, Ovstebo R, et al. Identification of meningococcal LPS as a major monocyte activator in IL-10 depleted shock plasmas and CSF by blocking the CD14-TLR4 receptor complex. J Endotoxin Res. 2003;9(3):155–63.
28. Moller AS, Ovstebo R, Haug KB, et al. Chemokine production and pattern recognition receptor (PRR) expression in whole blood stimulated with pathogen-associated molecular patterns (PAMPs). Cytokine. 2005;32(6):304–15.
29. Mogensen TH, Paludan SR, Kilian M, et al. Live *Streptococcus pneumoniae, Haemophilus influenzae*, and *Neisseria meningitidis* activate the inflammatory response through Toll-like receptors 2, 4, and 9 in species-specific patterns. J Leuk Biol. 2006;80(2):267–77.
30. Peiser L, De Winther MP, Makepeace K, et al. The class A macrophage scavenger receptor is a major pattern recognition receptor for *Neisseria meningitidis* which is independent of lipopolysaccharide and not required for secretory responses. Infect Immun. 2002;70(10):5346–54.
31. Constantin D, Cordenier A, Robinson K, et al. *Neisseria meningitidis*-induced death of cerebrovascular endothelium: mechanisms triggering transcriptional activation of inducible nitric oxide synthase. J Neurochem. 2004;89(5):1166–74.
32. Deghmane AE, El Kafsi H, Giorgini D, et al. Late repression of NF-kappaB activity by invasive but not non-invasive meningococcal isolates is required to display apoptosis of epithelial cells. PLoS Pathog. 2011;7(12):e1002403.
33. Chen Y, Wermeling F, Sundqvist J, et al. A regulatory role for macrophage class A scavenger receptors in TLR4-mediated LPS responses. Eur J Immunol. 2010;40(5):1451–60.
34. Gopinathan U, Brusletto BS, Olstad OK, et al. IL-10 immunodepletion from meningococcal sepsis plasma induces extensive changes in gene expression and cytokine release in stimulated human monocytes. Innate Immun. 2015;21(4):429–49.
35. Lu M, Zhang M, Takashima A, et al. Lipopolysaccharide deacylation by an endogenous lipase controls innate antibody responses to Gram-negative bacteria. Nat Immunol. 2005;6(10):989–94.
36. Mukhopadhyay S, Pluddemann A, Hoe JC, et al. Immune inhibitory ligand CD200 induction by TLRs and NLRs limits macrophage activation to protect the host from meningococcal septicemia. Cell Host Microbe. 2010;8(3):236–47.
37. Fransen F, Hamstra HJ, Boog CJ, et al. The structure of *Neisseria meningitidis* lipid A determines outcome in experimental meningococcal disease. Infect Immun. 2010;78(7):3177–86.
38. Read RC, Pullin J, Gregory S, et al. A functional polymorphism of toll-like receptor 4 is not associated with likelihood or severity of meningococcal disease. J Infect Dis. 2001;184(5):640–2.
39. Allen A, Obaro S, Bojang K, et al. Variation in Toll-like receptor 4 and susceptibility to group A meningococcal meningitis in Gambian children. Pediatr Infect Dis J. 2003;22(11):1018–9.

40. Gowin E, Swiatek-Koscielna B, Kaluzna E, et al. Analysis of TLR2, TLR4, and TLR9 single nucleotide polymorphisms in children with bacterial meningitis and their healthy family members. Int J Infect Dis. 2017;60:23–8.
41. Biebl A, Muendlein A, Kazakbaeva Z, et al. CD14 C-159T and toll-like receptor 4 Asp299Gly polymorphisms in surviving meningococcal disease patients. PLoS One. 2009;4(10):e7374.
42. Telleria-Orriols JJ, Garcia-Salido A, Varillas D, et al. TLR2-TLR4/CD14 polymorphisms and predisposition to severe invasive infections by *Neisseria meningitidis* and *Streptococcus pneumoniae*. Med Intensiva. 2014;38(6):356–62.
43. Erridge C, Stewart J, Poxton IR. Monocytes heterozygous for the Asp299Gly and Thr399Ile mutations in the Toll-like receptor 4 gene show no deficit in lipopolysaccharide signalling. J Exp Med. 2003;197(12):1787–91.
44. Esposito S, Bosis S, Orenti A, et al. Genetic polymorphisms and the development of invasive bacterial infections in children. Int J Immunopathol Pharmacol. 2016;29(1):99–104.
45. Mogensen TH, Berg RS, Paludan SR, et al. Mechanisms of dexamethasone-mediated inhibition of Toll-like receptor signaling induced by *Neisseria meningitidis* and *Streptococcus pneumoniae*. Infect Immun. 2008;76(1):189–97.
46. Opal SM, Laterre PF, Francois B, et al. Effect of eritoran, an antagonist of MD2-TLR4, on mortality in patients with severe sepsis: the ACCESS randomized trial. JAMA. 2013;309(11):1154–62.
47. Levin M, Quint PA, Goldstein B, et al. Recombinant bactericidal/permeability-increasing protein (rBPI(21)) as adjunctive treatment for children with severe meningococcal sepsis: a randomised trial. Lancet. 2000;356(9234):961–7.
48. Sprong T, Ley P, Abdollahi-Roodsaz S, et al. *Neisseria meningitidis* lipid A mutant LPSs function as LPS antagonists in humans by inhibiting TLR 4-dependent cytokine production. Innate Immun. 2011;17(6):517–25.
49. Jemmett K, Macagno A, Molteni M, et al. A cyanobacterial LPS-antagonist inhibits cytokine production induced by *Neisseria meningitidis* infection of a human whole blood model of septicaemia. Infect Immun. 2008;76:3156–63.
50. Piazza M, Rossini C, Della Fiorentina S, et al. Glycolipids and benzylammonium lipids as novel antisepsis agents: synthesis and biological characterization. J Med Chem. 2009;52(4):1209–13.
51. Brandtzaeg P. Host response to *Neisseria meningitidis* lacking lipopolysaccharides. Expert Rev Anti-Infect Ther. 2003;1(4):589–96.
52. Al-Bader T, Christodoulides M, Heckels JE, et al. Activation of human dendritic cells is modulated by components of the outer membranes of *Neisseria meningitidis*. Infect Immun. 2003;71:5590–7.
53. Rodriguez T, Perez O, Menager N, et al. Interactions of proteoliposomes from serogroup B *Neisseria meningitidis* with bone marrow-derived dendritic cells and macrophages: adjuvant effects and antigen delivery. Vaccine. 2005;23(10):1312–21.
54. Humphries HE, Triantafilou M, Makepeace BL, et al. Activation of human meningeal cells is modulated by lipopolysaccharide (LPS) and non-LPS components of *Neisseria meningitidis* and is independent of Toll-like receptor (TLR)4 and TLR2 signalling. Cell Microbiol. 2005;7:415–30.
55. Herbst-Kralovetz MM, Quayle AJ, Ficarra M, et al. Quantification and comparison of toll-like receptor expression and responsiveness in primary and immortalized human female lower genital tract epithelia. Am J Rep Immunol (New York, NY: 1989). 2008;59(3):212–24.
56. Laniewski P, Gomez A, Hire G, et al. Human three-dimensional endometrial epithelial cell model to study host interactions with vaginal bacteria and *Neisseria gonorrhoeae*. Infect Immun. 2017;85(3):pii: e01049-16.
57. Fichorova RN, Cronin AO, Lien E, et al. Response to *Neisseria gonorrhoeae* by cervicovaginal epithelial cells occurs in the absence of toll-like receptor 4-mediated signaling. J Immunol. 2002;168(5):2424–32.
58. Muenzner P, Naumann M, Meyer TF, et al. Pathogenic *Neisseria* trigger expression of their carcinoembryonic antigen-related cellular adhesion molecule 1 (CEACAM1; previously

CD66a) receptor on primary endothelial cells by activating the immediate early response transcription factor, nuclear factor-kappaB. J Biol Chem. 2001;276(26):24331–40.
59. Sintsova A, Guo CX, Sarantis H, et al. Bcl10 synergistically links CEACAM3 and TLR-dependent inflammatory signalling. Cell Microbiol. 2018;20(1):e12788.
60. Kalantari P, Harandi OF, Agarwal S, et al. miR-718 represses proinflammatory cytokine production through targeting phosphatase and tensin homolog (PTEN). J Biol Chem. 2017;292(14):5634–44.
61. Packiam M, Wu H, Veit SJ, et al. Protective role of Toll-like receptor 4 in experimental gonococcal infection of female mice. Mucosal Immunol. 2012;5(1):19–29.
62. Packiam M, Yedery RD, Begum AA, et al. Phosphoethanolamine decoration of *Neisseria gonorrhoeae* lipid A plays a dual immunostimulatory and protective role during experimental genital tract infection. Infect Immun. 2014;82(6):2170–9.
63. Liu M, John CM, Jarvis GA. Induction of endotoxin tolerance by pathogenic *Neisseria* is correlated with the inflammatory potential of lipooligosaccharides and regulated by microRNA-146a. J Immunol. 2014;192(4):1768–77.
64. Ding J, Rapista A, Teleshova N, et al. *Neisseria gonorrhoeae* enhances HIV-1 infection of primary resting CD4+ T cells through TLR2 activation. J Immunol. 2010;184(6):2814–24.
65. Malott RJ, Keller BO, Gaudet RG, et al. *Neisseria gonorrhoeae*-derived heptose elicits an innate immune response and drives HIV-1 expression. Proc Natl Acad Sci U S A. 2013;110(25):10234–9.
66. Dobson-Belaire WN, Rebbapragada A, Malott RJ, et al. *Neisseria gonorrhoeae* effectively blocks HIV-1 replication by eliciting a potent TLR9-dependent interferon-alpha response from plasmacytoid dendritic cells. Cell Microbiol. 2010;12(12):1703–17.
67. Andrade WA, Agarwal S, Mo S, et al. Type I interferon induction by *Neisseria gonorrhoeae*: dual requirement of cyclic GMP-AMP synthase and Toll-like receptor 4. Cell Rep. 2016;15(11):2438–48.
68. Christodoulides M, Heckels J. Novel approaches to *Neisseria meningitidis* vaccine design. Pathog Dis. 2017;73:ftx033.
69. Fransen F, Stenger RM, Poelen MC, et al. Differential effect of TLR2 and TLR4 on the immune response after immunization with a vaccine against *Neisseria meningitidis* or *Bordetella pertussis*. PLoS One. 2010;5(12):e15692.
70. Steeghs L, den Hartog R, den Boer A, et al. Meningitis bacterium is viable without endotoxin. Nature. 1998;392(6675):449–50.
71. Steeghs L, Keestra AM, van Mourik A, et al. Differential activation of human and mouse Toll-like receptor 4 by the adjuvant candidate LpxL1 of *Neisseria meningitidis*. Infect Immun. 2008;76(8):3801–7.
72. van der Ley P, van den Dobbelsteen G. Next-generation outer membrane vesicle vaccines against *Neisseria meningitidis* based on nontoxic LPS mutants. Hum Vaccin. 2011;7(8):886–90.
73. Zughaier S, Steeghs L, van der Ley P, et al. TLR4-dependent adjuvant activity of *Neisseria meningitidis* lipid A. Vaccine. 2007;25(22):4401–9.
74. Zariri A, Pupo E, van Riet E, et al. Modulating endotoxin activity by combinatorial bioengineering of meningococcal lipopolysaccharide. Sci Rep. 2016;6:36575.
75. Fransen F, Boog CJ, van Putten JP, et al. Agonists of Toll-like receptors 3, 4, 7, and 9 are candidates for use as adjuvants in an outer membrane vaccine against *Neisseria meningitidis* serogroup B. Infect Immun. 2007;75(12):5939–46.
76. Mistretta N, Guy B, Berard Y, et al. Improvement of immunogenicity of meningococcal lipooligosaccharide by coformulation with lipidated transferrin-binding protein B in liposomes: implications for vaccine development. Clin Vaccine Immunol. 2012;19(5):711–22.

Toll-Like Receptor 4 Activation by Damage-Associated Molecular Patterns (DAMPs)

Monica Molteni and Carlo Rossetti

Abstract Mammalian toll-like receptors (TLRs) act as key sensors of pathogen-associated molecular patterns (PAMP), such as bacterial lipopolysaccharides (LPS), lipopeptides, and flagellin, which are molecular structures present in microbial cells but not in host cells. It was therefore considered that the TLRs play a central role in the discrimination between "self" and "non-self." However, since the discovery of their microbial ligands, many studies have shown that molecules derived from the host can act as TLR4 agonists. These endogenous TLR4 ligands tend to fall into the categories of released intracellular proteins, extracellular matrix (ECM) components, oxidatively modified lipids, and other soluble mediators. This review summarizes the evidence supporting the intrinsic capacity of TLR4 stimulation in some of these proposed endogenous ligands and discuss their mechanism of action, often different from direct TLR4/MD-2 binding and canonical agonism.

Keywords TLR4 activation · NF-kB · Innate immunity · Stem cells · Tissue regeneration

1 Introduction

Toll-like receptors (TLRs) are receptors capable of detecting minute amounts of molecules released from microbial cells such as lipopolysaccharides (LPS), lipopeptides, and flagellin, and these molecules are generically called pathogen-associated molecular patterns (PAMPs). For this reason, TLRs have been considered to accomplish the exclusive task of recognizing exogenous molecules and to be involved in the first step of innate immune response activation by pathogens.

M. Molteni · C. Rossetti (✉)
Department of Medicine and Surgery, University of Insubria, Varese, Italy
e-mail: carlo.rossetti@uninsubria.it

C. Rossetti, F. Peri (eds.), *The Role of Toll-Like Receptor 4 in Infectious and Non Infectious Inflammation*, Progress in Inflammation Research 87,
https://doi.org/10.1007/978-3-030-56319-6_6

However, the observation that nonmicrobial ligands, linked to tissue and cellular damage, can activate TLRs led to the conclusion that even molecules originated from the host can activate innate immunity. These endogenous TLR activators are generally called Damage-Associated Molecular Patterns (DAMPs), suggesting that the main function of innate immune activation by TLRs is not only to distinguish the self from the nonself, but also to contrast the damage [1–6].

The TLR that recognizes gram-negative bacteria LPS is TLR4, and it is expressed in the cell membranes of circulating cells belonging to the immune system and of some cells situated in peripheral tissues [7–12]. TLR4 activation by bacterial endotoxin (LPS and lipooligosaccharide (LOS)) occurs by the cooperation of TLR4 and other extracellular LPS-binding proteins (CD14, MD-2) and leads to the induction of an intracellular signaling cascade, regardless of both PAMP and DAMP molecules, resulting in an activation of transcriptional factors, including NF-κB [51].

NF-κB is one of the main transcriptional factors in the activation of genes involved in innate immune triggering and coding for pro-inflammatory cytokines [26, 32]. Host necrotic cells as well can act as strong activators of NF-κB inducing a strong pro-inflammatory response [13, 14]. Conversely, apoptotic cells do not induce the expression of pro-inflammatory cytokines because they are not able to significantly activate NF-κB. Downstream pathophysiological effects due to NF-κB activation induced by necrotic cellular death encompass: activation of genes coding for pro-inflammatory cytokines and of genes involved in the process of repairing and remodeling of tissues, transmission of the distress signal to neighboring cells, activation of inflammatory process and innate immune cells to resolve the damage, activation of cellular networks to instruct adaptive immunity and establish immunological memory, and the management of the reparation and the recovery of tissues, based on the interaction between different DAMPs.

While the activation of NK-κB is common for resident tissue cells and professional immune cells, the main difference caused by the activation of TLR4 seems to be that in immune cells TLR4 triggering is needed to activate the complex immune response against antigens. On the other hand, TLR4 triggering on resident tissue cells is primarily needed for the local recruitment of specialized immune cells and cells that have a key role in tissue recovery (polymorphonuclear phagocytes, monocyte/macrophages, fibrocytes, and tissue stem cells).

TLR4 is expressed also on several subsets of stem/progenitor cells (SCs). In these cells, the role of TLR4 is related to the regulation of the basal motility, proliferation, differentiation processes, self-renewal, and immunomodulation [25]. Increasing the knowledge about the involvement of TLRs in the response of SCs to specific tissue damage and in the reparative processes could be decisive for the development of new therapeutic strategies in the tissue recovery.

Besides their beneficial role in the response to tissue damage, DAMPs can contribute to the pathogenesis of different inflammatory and autoimmune diseases, which are characterized by an aberrant TLR activation [26]. Indeed, it is more likely that DAMPs could work as the amplifiers of these pathologies, rather than being the cause of them. DAMPs are not only released by dead cells, but can be secreted by living cells that are subjected to stress. DAMPs are directly connected to inflammation and related disorders: In fact, the inhibition of immune responses mediated by DAMPs is a promising strategy to improve the clinical management of inflammatory diseases caused by infections and wounds. However, it is important to keep in mind that DAMPs are not only distress signals, but also play a main role in tissues repairing. Some DAMPs have been studied for their role in the healing process of tissues after sterile inflammation or infectious inflammatory response [15–23].

2 Damage-Associated Molecular Patterns (DAMPs)

In recent years, several molecules released by tissue and cellular damage have been identified as agonists of TLR4. Some of these molecules are situated inside of the cells that are normally not accessible from immune system, but are released outside as a consequence of cells death and the breaking of plasma membrane. Some other molecules are linked to the damage of extracellular matrix, and are in the majority of cases low-molecular-weight fragments derived from extracellular matrix molecules released after tissue damage.

One of the first reports describing the involvement of TLR4 and DAMPs was published in 2000. According to this paper, heat shock protein 60 (HSP60) is able to induce the production of pro-inflammatory cytokines through the activation of TLR4 [12].

From then, endogenous ligands able to activate TLR4 intracellular signaling have increased and include molecules belonging to HSP family [12, 15–19, 26–28, 39], Gp96 (heat shock protein 90 kDa beta family of calcium-binding cytosolic proteins) [12, 16, 38], high-mobility group box 1 (HMGB1) [3, 30, 33], and S100 calcium-binding protein (a family of calcium-binding cytosolic proteins) [46–48], representing intracellular DAMPs. As tissue DAMPs, oligosaccharides [36], biglycans [52, 53], and tenascin C [54, 55] of hyaluronic acid have been demonstrated to activate TLR4.

Given the heterogeneity of DAMPs molecules and the fact that some other accessory molecules take part in this process to make the receptor work, it is not possible to speculate only on a direct action mediated by agonist activity on TLR4. At least three other possible mechanisms have been supposed, other than TLR direct agonism:

1. An action mediated by other receptors with other intracellular pathways that can modulate the "intracellular signaling" of TLRs and in particular of TLR4.
2. An effect of implementation in the presentation of PAMPs by DAMPs (chaperon effect), molecule–molecule interaction.
3. An action on co-signaling accessory molecules, in particular CD14.

3 Heat Shock Proteins (HSPs)

HSPs are involved in the interaction between innate and adaptive immune system and mainly mediate immune regulatory functions. HSPs represent a combination of evolutionarily conserved proteins, molecular chaperones, which bind to some other proteins in a nonnative form and help them to reach the functional conformation. They have been described also as molecular shuttles of antigens, especially in tumor cells [17–19, 22, 24, 26].

The mechanism of action seems to be linked to a close connection established between HSPs and PAMPs; further studies have demonstrated that HSP60 can bind with high affinity to LPS in a saturable way and strengthen its biological activity with a synergistic effect [26].

Recent remarks show that endogenous distress signals and microbial products can cooperate in the induction of innate immunity directly working both on the presentation of molecules, and on the activation of other receptors that cooperate with TLR4 to activate intracellular signaling pathways [27].

Interactions between DAMPs and PAMPs and their receptors, which are temporary and spatially overlapped, may be the molecular basis for the observation that infections, just like nonspecific triggers, could be at the basis of rheumatic diseases. It has been suggested that the synergy between HSPs and PAMPs could be responsible for high and premature immune responses during infections [28, 29, 40, 42, 44].

4 Gp96

Gp96 is a type of HSP with different functions as DAMP: one inside the cell as a chaperone of TLR4 intracellular signaling, and one outside the cell as a chaperone both of TLR2 and TLR4. The extracellular interaction seems to involve the N-terminal domains [12, 16, 23, 45].

Experiments using Gp96 or its N-terminal domain, nominally lacking endotoxins (<0,5 enzymatic units/mg), showed cellular innate immune activation only at high concentrations (≥ 50 μg/ml). However, the preincubation of dendritic cells with low amounts of Gp96, at a concentration that makes it unable to activate dendritic cells, can induce higher levels of pro-inflammatory cytokines and activation markers after PAMPs incubation compared to dendritic cells stimulated with PAMPs alone, inducing also the activation of T-cells. These results provide new significant information about the mechanism of action HSPs in dendritic cells, showing an important effect of HSPs as amplifiers of the danger signals originated from exogenous bacteria and as activators of adaptive immune responses [41].

5 HMGB1

High-mobility group box 1 (HMGB1) is a non-histone nuclear protein that acts as an alarmin, inducing a proinflammatory response, and has been found to be involved in the pathogenesis of inflammatory and autoimmune diseases. It has been demonstrated that during cell activation and cell death, HMGB1 can translocate from the nucleus to the cytoplasm and finally be released in the extracellular environment. HMGB1 binds TLR4, and this interaction has been found to be necessary for cytokine induction; TLR4-deficient macrophages do not release proinflammatory cytokines in response to HMGB1. Activity of HMGB1 is dependent on the redox state of its cysteine residues. In fact, the recruitment of inflammatory cells via TLR4 has been demonstrated to be mediated by the reduced form of CyS106 in HMGB1. It has been hypothesized that modification of HMGB1 redox state might contribute to the different immune response observed in the presence of necrotic or apoptotic cells. Necrotic cells release the reduced form of HMGB1 and can activate inflammation, whereas apoptotic cells release HMGB1 with Cys106 in an oxidized state, so they are unable to stimulate TLR4 and, therefore, they cannot induce a pro-inflammatory response [3, 30]. In addition to the role of a thiol group at Cys106, TLR4 stimulation by HMGB1 also needs a concomitant disulfide Cys23–Cys45 linkage.

Independently of its intrinsic cytokine-inducing ability, HMGB1 can promote inflammation by forming immunostimulatory complexes with other mediators, such as IL-1, lipopolysaccharides, and DNA. These complexes have been identified by means of immunoprecipitation and co-stimulation assays. It has been shown that complexes can dramatically enhance cell responses when compared with induction by the ligand alone. However, some evidence suggests that HMGB1–partner molecule complexes signal through the receptor of the partner molecule, even though the mechanism is not clear. Limited information is available about the nature of the structures formed, and the kinetics of complex formation. Some studies have shown that in vitro interaction of HMGB1 with lipopolysaccharide is time and concentration-dependent and can occur over many hours. These findings also suggest that conformational changes in the partner molecules are required for a stable interaction.

In addition to effects on immune cells, HMGB1 can modulate the activities of other cells, including hematopoietic, epithelial, and neuronal cells and mediate systemic effects such as fever, anorexia, and acute-phase responses. These diverse activities reflect its function as an alarmin and depends on the ability of HMGB1 to engage several receptors including TLR2, TLR4, TLR9, receptor for advanced glycation end products (RAGE), and CD24 [3, 30, 31, 33, 34]. In particular, RAGE was the first demonstrated HMGB1-binding partner [32]. RAGE is a transmembrane, multiligand member of the immunoglobulin superfamily. As shown in vitro, HMGB1 signaling through RAGE mediates: chemotaxis, proliferation and differentiation of immune and nonimmune cells, and upregulation of cell-surface receptors, among which are TLR4 and RAGE.

6 S100 Calcium Binding Proteins (Family of Calcium-Binding Cytosolic Proteins)

S100 cytoplasmic proteins belong to the family of cytosolic proteins that bind calcium, are produced by myeloid cells, and are promising markers of inflammation. S100A8/A9 and S100A12 are released by monocytes and granulocytes during the activation of innate immune system. Tissue and serum concentrations of S100 proteins are linked to the activity of diseases, both during local inflammation and systemic inflammation. It has been found that in diseases, such as familial Mediterranean fever (FMF) and systemic juvenile idiopathic arthritis (sJIA), there is a hypersecretion of S100 proteins [46]. Extracellular S100 proteins, working as distress signals, are crucial in the regulation of immune homeostasis also in posttraumatic injuries, through the interaction with TLR4 [48]. The dimer seems to be able to bind to TLR4–MD2 complex through the c-terminal domain, leading to its activation. Recently, it has been outlined that, in an animal model of renal ischemia/reperfusion (I/R), S100A8/A9 do not contribute to the acute injury induced by I/R [46]. However, using S100A9 KO mice, an improvement in renal dysfunction, in the extent of the damage and reduction of fibrosis and inflammation, phenomena all associated with an advanced polarization of subtype M2 macrophages, has been observed [49].

7 Hyaluronic Acid (HA)

HA is a main constituent of the extracellular matrix and has a main role in the process of tissue renewal. As constituent of the extracellular matrix, it is deemed to be involved in maintaining tissues hydrated, distributing and transporting plasmatic proteins, and in maintaining the matrix structure undamaged. It has been demonstrated that it can control inflammation, cellular migration, and angiogenesis, which have specific roles in the main phases of wound healing, through specific receptors.

HA is a linear polysaccharide with high molecular weight (~ 800 KDa), made of repeating disaccharide units of β1-3 linked D-glucuronic acid (GlcA), and residues of β1-3 linked N-acetyl-D-glucosamine (GlcNAc) (4GlcAβ1 -3GlcNAc1β).

Some papers have outlined that most of the properties of HA depend on its molecular weight [37, 50]. A high molecular weight of HA features anti-inflammatory and immunosuppressive properties, while HA at a low molecular weight shows a pro-inflammatory activity. Recent studies have demonstrated that fragments of HA can be used as an endogenous trigger of TLR4 signaling, and can induce innate immunity, promoting the production of cytokines from different cellular types. Fragments of HA are produced during inflammation or injury through the activity of oxygen's radicals or through enzymatic activity by hyaluronidases, β-glucuronidases, and hexosaminidases.

It is been shown that CD44, a HA receptor, is involved in the release of pro-inflammatory cytokines and is induced by HA to form a TLR4-CD44 complex [36].

8 Biglycans

Biglycans are small proteoglycans rich in leucine, present in different extracellular matrixes such as bones, cartilage, tendons, etc. However, their biological role is not clear yet.

It has proved that biglycans work in macrophages as an endogenous ligand of TLR4, leading to a rapid activation of p38, ERK, and NF-κB [35, 43, 52].

Recently, several authors have observed that biglycans work as an extracellular distress signal derived from the matrix in its soluble form. It can act as a ligand with high affinity for TLR4 co-receptors, particularly CD14 and CD44, providing a further complexity level in the regulation of TLR4 activation. The specific interaction of biglycans with CD14 is important for TLR4 triggering, while by binding with CD44 it induces autophagy [53].

Biglycans, and maybe some other soluble proteoglycans, may work as molecular switches that could either propagate the inflammation signal improving the interaction CD14-LPS or promote the resolution of the inflammatory process, activating the autophagy. These new functions can have important consequences in the regulation of different inflammatory diseases and could provide the basis for the development of new therapeutic approaches.

9 Tenascin-C

Tenascin-C is a glycoprotein of the extracellular matrix involved in inflammation and tissue damage, mainly in rheumatoid arthritis. Mice that do not express tenascin-C show a rapid resolution of acute articular inflammation and are protected by erosive arthritis [54]. Intra-articular injection of tenascin-C promotes inflammation in vivo and the incubation of synovial cells from patients with rheumatoid arthritis with exogenous tenascin-C induces the secretion of pro-inflammatory cytokines [55]. Furthermore, in human macrophages and fibroblasts isolated from synovium of patients with rheumatoid arthritis, tenascin-C induces the synthesis of pro-inflammatory cytokines through the activation of TLR4, thus suggesting tenascin-C as a new endogenous activator of TLR4-mediated innate immune response, controlling synovial persistent inflammation and tissue joint destruction. Moreover, other authors have provided data that show how tenascin-C, after cellular stimulation with LPS, sustains the synthesis of TNF-α controlling the expression of miR-155 [55, 56].

Table 1 List of some TLR4 DAMPs

Origin	DAMPs	Receptors
Intracellular compartment		
	Nuclear Histones	TLR2, TLR4,
	HMGB1	TLR2, TLR4, RAGE
	Defensins	TLR4
	Granulysin	TLR4
Extracellular compartment		
	Biglycans	NLRP3, TLR4, TLR2
	LMW Hyaluronan	TLR2, TLR4, NLRP3
	Fibronectin (EDA domain)	TLR4
	Fibronogen	TLR4
	Tenascin C	TLR4
	S100 proteins	TLR2, TLR4, RAGE

10 Conclusion

It is clear that the role of TLR4 cannot be seen only as an important receptor in the activation of innate immune response due to pathogens but can also be seen as a receptor activated by host cell and tissue damage. The most recent observations show that stem cells express TLR4; this indicates a more complex role of this receptor in the mechanisms of tissue recovery after damage. It is also important to highlight that the activity of DAMPS may not be due to a direct agonist action on the receptor but rather they act as a co-stimulatory signal on TLR4. This observation could explain the involvement of TLR4 in some autoimmune diseases characterized by tissue destruction, such as rheumatoid arthritis (Table 1).

References

1. Matzinger P. Tolerance, danger, and the extended family. Annu Rev Immunol. 1994;12:991–1045.
2. Beutler B. Neo-ligands for innate immune receptors and the etiology of sterile inflammatory disease. Immunol Rev. 2007;220(1):113–28.
3. Bianchi ME. DAMPs, PAMPs and alarmins: all we need to know about danger. J Leukoc Biol. 2007;81(1):1–5.
4. Gordon S. Pattern recognition receptors: doubling up for the innate immune response. Cell. 2002;111(7):927–30.
5. Medzhitov R, Janeway CA Jr. Decoding the patterns of self and nonself by the innate immune system. Science. 2002;296(5566):298–300.
6. Kono H, Rock KL. How dying cells alert the immune system to danger. Nat Rev Immunol. 2008;8(4):279–89.
7. Chen K, Huang J, Gong P, Iribarren NMD, Wang JM. Toll-like receptors in inflammation, infection and cancer. Int Immunopharmacol. 2007;7(10):1271–85.

8. Liew FY, Xu D, Brint EK, O'Neill LAJ. Negative regulation of Toll-like receptor-mediated immune responses. Nat Rev Immunol. 2005;5(6):446–58.
9. Lotze MT, Zeh HJ, Rubartelli A, et al. The grateful dead damage-associated molecular pattern molecules and reduction/oxidation regulate immunity. Immunol Rev. 2007;220(1):60–81.
10. Mantovani A. Cancer: inflaming metastasis. Nature. 2009;457(7225):36–7.
11. Midwood KS, Piccinini AM, Sacre S. Targeting toll-like receptors in autoimmunity. Curr Drug Targets. 2009;10(11):1139–55.
12. Ohashi K, Burkart V, Flohé S, Kolb H. Cutting edge: heat shock protein 60 is a putative endogenous ligand of the Toll-like receptor-4 complex. J Immunol. 2000;164(2):558–61.
13. Basu S, Binder RJ, Suto R, Anderson KM, Srivastava PK. Necrotic but not apoptotic cell death releases Mediators of Inflammation 13 heat shock proteins, which deliver a partial maturation signal to dendritic cells and activate the NF-κB pathway. Int Immunol. 2000;12(11):1539–46.
14. Li M, Carpio DF, Zheng Y, et al. An essential role of the NF-κB/Toll-like receptor pathway in induction of inflammatory and tissue-repair gene expression by necrotic cells. J Immunol. 2001;166(12):7128–35.
15. Asea M, Rehli M, Kabingu E, et al. Novel signal transduction pathway utilized by extracellular HSP70: role of Toll-like receptor (TLR) 2 and TLR4. J Biol Chem. 2002;277(17):15028–34.
16. Ohashi K, Burkart V, Flohé S, Kolb H. Cutting edge: heat shock protein 60 is a putative endogenous ligand of the Toll-like receptor-4 complex. J Immunol. 2000;164(2):558–61.
17. Roelofs MF, Boelens WC, Joosten LAB, et al. Identification of small heat shock protein B8 (HSP22) as a novel TLR4 ligand and potential involvement in the pathogenesis of rheumatoid arthritis. J Immunol. 2006;176(11):7021–7.
18. Wheeler DS, Chase MA, Senft AP, Poynter SE, Wong HR, Page K. Extracellular Hsp72, an endogenous DAMP, is released by virally infected airway epithelial cells and activates neutrophils via Toll-like receptor (TLR)-4. Respir Res. 2009;10:3.
19. Asea A, Rehli M, Kabingu E, et al. Novel signal transduction pathway utilized by extracellular HSP70: role of Toll-like receptor (TLR) 2 and TLR4. J Biol Chem. 2002;277(17):15028–34.
20. Piccinini AM, Midwood KS. DAMPening inflammation by modulating TLR signalling. Mediat Inflamm. 2010;2010:1–21.
21. Zhang B, Du Y, He Y, Liu Y, Zhang G, Yang C, Gao F. INT-HA induces M2-like macrophage differentiation of human monocytes via TLR4-miR-935 pathway. Cancer Immunol Immunother. 2019;68(2):189–200.
22. Vabulas RM, Ahmad-Nejad P, Ghose S, Kirschning CJ, Issels RD, Wagner H. HSP70 as endogenous stimulus of the toll/interleukin-1 receptor signal pathway. J Biol Chem. 2002;277(17):15107–12.
23. Vabulas RM, Braedel S, Hilf N, et al. The endoplasmic reticulum-resident heat shock protein Gp96 activates dendritic cells via the Toll-like receptor 2/4 pathway. J Biol Chem. 2002;277(23):20847–53.
24. Roelofs MF, Boelens WC, Joosten LAB, et al. Identification of small heat shock protein B8 (HSP22) as a novel TLR4 ligand and potential involvement in the pathogenesis of rheumatoid arthritis. J Immunol. 2006;176(11):7021–7.
25. Sallustio F, Curci C, Stasi A, De Palma G, Divella C, Gramignoli R, Castellano G, Gallone A, Gesualdo L. Role of Toll-like receptors in actuating stem/progenitor cell repair mechanisms: different functions in different cells. Stem Cells Int. 2019;2:1–12.
26. Chang ZL. Important aspects of Toll-like receptors, ligands and their signaling pathways. Inflamm Res. 2010;59(10):791–808.
27. Wheeler DS, Chase MA, Senft AP, Poynter SE, Wong HR, Page K. Extracellular Hsp72, an endogenous DAMP, is released by virally infected airway epithelial cells and activates neutrophils , via toll-like receptor (TLR)-4. Respir Res. 2009;10:31.
28. Kol A, Lichtman AH, Finberg RW, Libby P, Kurt-Jones EA. Cutting edge: heat shock protein (HSP) 60 activates the innate immune response: CD14 is an essential receptor for HSP60 activation of mononuclear cells. J Immunol. 2000;164(1):13–7.

29. Libby P, Kurt-Jones EA. Cutting edge: heat shock protein (HSP) 60 activates the innate immune response: CD14 is an essential receptor for HSP60 activation of mononuclear cells. J Immunol. 2000;164(1):13–7.
30. Abreu MT, Arditi M. Innate immunity and toll-like receptors: clinical implications of basic science research. J Pediatr. 2004;144(4):421–9.
31. Park JS, Svetkauskaite D, He Q, et al. Involvement of Toll-like receptors 2 and 4 in cellular activation by high mobility group box 1 protein. J Biol Chem. 2004;279(9):7370–7.
32. Klune JR, Dhupar R, Cardinal J, Billiar TR, Tsung A. HMGB1: endogenous danger signaling. Mol Med. 2008;14(7–8):476–84.
33. Ghosh S, May MJ, Kopp EB. NF-κB and Rel proteins: evolutionarily conserved mediators of immune responses. Annu Rev Immunol. 1998;16:225–60.
34. Park JS, Gamboni-Robertson F, He Q, et al. High mobility group box 1 protein interacts with multiple Toll-like receptors. Am J Phys. 2006;290(3):C917–24.
35. Schaefer L, Babelova A, Kiss E, et al. The matrix component biglycan is proinflammatory and signals through Toll-like receptors 4 and 2 in macrophages. J Clin Investig. 2005;115(8):2223–33.
36. Midwood K, Sacre S, Piccinini AM, et al. Tenascin-C is an endogenous activator of Toll-like receptor 4 that is essential for maintaining inflammation in arthritic joint disease. Nat Med. 2009;15(7):774–80.
37. Termeer C, Benedix F, Sleeman J, et al. Oligosaccharides of hyaluronan activate dendritic cells via Toll-like receptor 4. J Exp Med. 2002;195(1):99–111.
38. Johnson GB, Brunn GJ, Kodaira Y, Platt JL. Receptor-mediated monitoring of tissue well-being via detection of soluble heparan sulfate by Toll-like receptor 4. J Immunol. 2002;168(10):5233–9.
39. Radsak MP, Hilf N, Singh-Jasuja H, et al. The heat shock protein Gp96 binds to human neutrophils and monocytes and stimulates effector functions. Blood. 2003;101(7):2810–5.
40. Osterloh A, Kalinke U, Weiss S, Fleischer B, Breloer M. Synergistic and differential modulation of immune responses by Hsp60 and lipopolysaccharide. J Biol Chem. 2007;282(7):4669–80.
41. Warger T, Hilf N, Rechtsteiner G, et al. Interaction of TLR2 and TLR4 ligands with the N-terminal domain of Gp96 amplifies innate and adaptive immune responses. J Biol Chem. 2006;281(32):22545–53.
42. Trompette A, Divanovic S, Visintin A, et al. Allergenicity resulting from functional mimicry of a Toll-like receptor complex protein. Nature. 2009;457(7229):585–8.
43. Babelova A, Moreth K, Tsalastra-Greul W, et al. Biglycan, a danger signal that activates the NLRP3 inflammasome via toll-like and P2X receptors. J Biol Chem. 2009;284(36):204035–48.
44. Vabulas RM, Ahmad-Nejad P, da Costa C, et al. Endocytosed HSP60s use toll-like receptor 2 (TLR2) and TLR4 to activate the toll/interleukin-1 receptor signaling pathway in innate immune cells. J Biol Chem. 2001;276(33):31332–9.
45. Vabulas RM, Braedel S, Hilf N, Singh-Jasuja H, Herter S, Ahmad-Nejad P, Kirschning CJ, Da Costa C, Rammensee HG, Wagner H, Schild H. The endoplasmic reticulum-resident heat shock protein Gp96 activates dendritic cells via the Toll-like receptor 2/4 pathway. J Biol Chem. 2002;277(23):20847–5.
46. Yang Y, Liu B, Dai J, Srivastava PK, Zammit DJ, Lefrançois L, Li Z. Heat shock protein gp96 is a master chaperone for toll-like receptors and is important in the innate function of macrophages. Immunity. 2007;26(2):215–26.
47. Dessing MC, Tammaro A, Pulskens WP, Teske GJ. The calcium-binding protein complex S100A8/A9 has a crucial role in controlling macrophage-mediated renal repair following ischemia/reperfusion. Kidney Int. 2015;87(1):85–94.
48. Donato R, Cannon BR, Sorci G, Riuzzi F, Hsu K, Weber DJ, Geczy CL. Functions of S100 proteins. Curr Mol Med. 2013;13(1):24–57.

49. Cerezo LA, Remáková M, Tomčik M, Gay S, Neidhart M, Lukanidin E, Pavelka K, Grigorian M, Vencovský J, Šenolt L. The metastasis-associated protein S100A4 promotes the inflammatory response of mononuclear cells via the TLR4 signalling pathway in rheumatoid arthritis. Rheumatology (Oxford). 2014;53(8):1520–6.
50. Voelcker V, Gebhardt C, Averbeck M, Saalbach A, Wolf V, Weih F, Sleeman J, Anderegg U, Simon J. Hyaluronan fragments induce cytokine and metalloprotease upregulation in human melanoma cells in part by signalling via TLR4. Exp Dermatol. 2008;17(2):100–7.
51. Lu YC, Yeh WC, Ohashi PS. LPS/TLR4 signal transduction pathway. Cytokine. 2008;42(2):145–51.
52. Johnson GB, Brunn GJ, Kodaira Y, Platt JL. Receptor-mediated monitoring of tissue well-being via detection of soluble heparan sulfate by Toll-like receptor 4. J Immunol. 2002;168(10):5233–9.
53. Schaefer L, Babelova A, Kiss E, Hausser HJ, Baliova M, Krzyzankova M, Marsche G, Young MF, Mihalik D, Götte M, Malle E, Schaefer RM, Gröne HJ. The matrix component biglycan is proinflammatory and signals through Toll-like receptors 4 and 2 in macrophages. J Clin Invest. 2005;115(8):2223–33.
54. Roedig H, Nastase MV, Wygrecka M, Schaefer L. Breaking down chronic inflammatory diseases: the role of biglycan in promoting a switch between inflammation and autophagy. FEBS J. 2019;286(15):2965–79.
55. Midwood KS, Orend G. The role of tenascin-C in tissue injury and tumorigenesis. J Cell Commun Signal. 2009;3(3–4):287–310.
56. Zuliani-Alvarez L, Marzeda AM, Deligne C, Schwenzer A, McCann FE, Marsden BD, Piccinini AM, Midwood KS. Mapping tenascin-C interaction with toll-like receptor 4 reveals a new subset of endogenous inflammatory triggers. Nat Commun. 2017;8(1):1595.

TLR4 in Neurodegenerative Diseases: Alzheimer's and Parkinson's Diseases

Claudia Balducci and Gianluigi Forloni

Abstract A role of inflammation in the pathogenesis of neurodegenerative disorders is commonly accepted. The excess of inflammatory factors production significantly contributes to neuronal dysfunction transforming the original protective function of glial cells in detrimental activities. This general mechanism is differently activated in each single disorder, and specific elements, biological pathways, and timing are combined to affect neuronal vulnerability. The delucidation of the sequence of events at the molecular level might indicate appropriate pharmacological tools to control inflammation with a potential therapeutic meaning. In this chapter, we analyze the role of toll-like receptor 4 (TLR4) as mediator of inflammatory response in Alzheimer's disease (AD) and Parkinson's disease (PD). In both pathological conditions the activation of TLR4 has been associated either to harmful or beneficial effects although the data supporting the negative consequence are prevailing. Numerous in vitro and in vivo studies indicate directly or indirectly interaction between the activation of TLR4 and the presence of β-amyloid. Accordingly, in our hands, the memory damage and inflammatory effects obtained by intraventricular application of β-amyloid oligomers were antagonized by TLR4 inhibitor and completely abolished in TLR4-knockout mice. Genetic studies associate missense mutation that attenuates the TLR4 signaling with a reduction of risk to develop AD. The beneficial effects of TLR4 in PD was shown in an experimental model where its ablation hampered the ability of microglia to phagocytize α-synuclein in vitro. In several other experimental conditions, the activation of TLR4 exacerbated the pathological outcomes, and the inhibition attenuated them. In contrast with the data obtained with β amyloid oligomers, the intraventricular application of α-synuclein oligomers was not influenced by TLR4 inhibition or ablation. A possible role of TLR4 has been proposed in the relationship between gut microbiota dysbiosis and increased risk of developing PD. In conclusion, the modulation of

C. Balducci · G. Forloni (✉)
Neuroscience Department, Istituto di Ricerche Farmacologiche Mario Negri IRCCS, Milan, Italy
e-mail: gianluigi.forloni@marionegri.it

C. Rossetti, F. Peri (eds.), *The Role of Toll-Like Receptor 4 in Infectious and Non Infectious Inflammation*, Progress in Inflammation Research 87,
https://doi.org/10.1007/978-3-030-56319-6_7

TLR4 is an interesting pharmacological target for both AD and PD and some data in these senses has been already shown; however, the relevance of physiological role of TLR4 must be carefully considered for any clinical application.

Keywords Neuroinflammation · Glial cells · Animal models of neurodegenerative diseases

1 Alzheimer's and Parkinson's Diseases

Alzheimer's and Parkinson's diseases (AD, PD) are the two most common neurodegenerative disorders affecting millions of people worldwide. AD and PD exist either in the familial form (5–10% of the cases), with a genetic origin, or in a sporadic one, which accounts for 90% of the cases. At neuropathological level, the two forms are indistinguishable, which suggests that common mechanisms are shared [1–3].

In terms of clinical symptoms, AD is mainly characterized by the presence of cognitive alterations such as loss of memory, attentional deficits, inability to execute daily live actions, disorientation, speech defects, etc. Neuronal loss occurs mainly in the cortical and hippocampal brain areas [4]. In PD, motor alterations are the primary symptoms including rigidity, asymmetric resting tremor, cogwheel rigidity, and bradykinesia. Nonmotor manifestations also occur, among which is cognitive dysfunction due to synaptic alterations, as in AD. Neuronal and terminal loss are detectable in the substantia nigra par compacta and in the striatum respectively [5].

AD and PD are mainly caused by the abnormal aggregation of misfolded proteins such as β-amyloid (Aβ) for AD and α-synuclein (αSyn) for PD [6]. Aβ is cleaved from the transmembrane amyloid precursor protein (APP), whereas αSyn [7, 8] is a natively unfolded cytoplasmic protein mostly expressed in neuronal presynaptic terminals and also detected in human fluids (i.e., plasma and cerebrospinal fluid) [9–11]. Both Aβ and αSyn enter an auto-aggregation process by which a series of well-organized polymers originate ranging from soluble, small species, namely the oligomers, up to intermediate and larger assembles termed protofibrils and fibrils, respectively. The fate of the larger, insoluble aggregates is either to deposit into the brain parenchyma as extracellular amyloid plaques in AD, which together with intracellular tau-enriched neurofibrillary tangles represent the two main histopathological hallmarks, and as intracellular Lewy bodies or Lewy bodies neuritis in PD. On the other hand, oligomeric forms freely circulate in the brain interacting with neurons and glial cells perturbing membrane permeability, neuronal and synaptic activity, and promoting neuroinflammation [12–14]. For these main reasons, the oligomers of misfolded proteins are nowadays recognized has the species with the most powerful multi-level toxicity, from which the term "oligomeroptahies" to classify these disorders [14]. In AD oligomers are the best correlates of disease severity and synaptic dysfunction, whereas the association between larger aggregates content and cell death has been ruled out [15–17]. Several experimental evidences support this statement by demonstrating that both Aβ and αSyn oligomer application directly on cultured neurons or in mouse brain could alter synaptic function and cognition, while larger aggregates are inactive [18–21].

Another significant detrimental effect of the oligomers is exerted on glial cells. Through a series of in vitro and in vivo studies it was shown that microglia and astrocytes exposed to the oligomers showed a stronger degree of activation compared to fibril exposure [22–25].

Neuroinflammation is one of the most significant neuropathological manifestation uniting both AD and PD [26–28]; also affected patients display activated glial cells and an increase in the levels of pro-inflammatory mediators, with interleukin 1β (IL-1β), tumors necrosis factor alpha (TNF-α), and interleukin 6 (IL-6) as the most representative [29]. Worth of note, the fact that despite neuroinflammation has been considered as a secondary event for many years, in the last years, it re-emerged as a driving force at the very early stages of disease as well as a core therapeutic target. Although glial cells, especially microglia, are deputed to an action of defense by continuously patrolling the brain microenvironment and eliminating noxious stimuli – among which misfolded proteins – when their engagement perpetuates such as in chronic pathological conditions, the continuous release of pro-inflammatory cytokines from these cells culminate into neurotoxic events leading to neuronal dysfunction and loss. In addition, different microglia activation states also entail a different control of synaptic activity and cognition. Indeed, microglia regulate and survey synaptic activity when in their "*resting, ramified state*"; alternatively, when they enter an "*activated, amoeboid state*," their processes retract from neurons and deprive them of that regulatory control necessary to guarantee new memory processing and consolidation [30].

Among the receptors implicated in neuroinflammatory events, Toll-like receptors 4 (TLR4) are considered as the first initiator of the innate immune response. However, despite numerous studies devoted to elucidating their involvement in neurodegenerative diseases, many controversies still exist. Through this chapter we will highlight all the evidence describing if and how TLR4 is implicated in AD and PD.

2 TLR4 in AD

TLR4 is a well-known family of pattern recognition receptors (PRRs) expressed on both neurons and glia under normal conditions [31], which seems to be involved in the pathogenetic cascade of AD. Neuroinflammation mostly occurs at the site of Aβ deposits, which are surrounded by activated microglia and astrocytes both expressing a series of immune receptors including TLRs [32]. TLR4 is important in the modulation of Aβ fibril clearance by microglial cells, a process associated with the activation of the inflammatory pathway involving the nuclear factor kappa-light chain-enhancer of activate B-cells, and the mitogen-activated protein kinase (NF-κB/MAPK). Main products of this innate immune pathway are pro-inflammatory cytokines such as IL-1β, TNF-α, and IL-6 all significantly expressed at higher level in the brain of AD patients and potentially responsible for neurotoxicity [33].

The role of TLR4 in AD is still extremely controversial, although most of the data favor TLR4's harmful effects over its benefits. Discrepancies originate from data showing that TLR4 expression is linked to Aβ uptake, indeed, as mentioned above, Aβ is scavenged by microglia through receptor-mediated phagocytosis and degradation, a process involving G-protein-coupled receptors (GPCRs), scavenger receptors, receptor for advance glycation end-product (RAGE), and TLRs especially TLR2, TLR4, and the co-receptor CD14 [34].

Protective TLR4 effects are suggested by the investigation on the role of TLR4 in relation to amyloidogenesis from Tahara and colleagues [34], who determined the amount of cerebral Aβ in AD mouse models with different genotypes of TLR4. They found that Mo/Hu APPswePS1dE9 mice (APP/PS1), homozygous for a destructive mutation of TLR4 [Tlr(Lps-d)/Tlr(Lps-d)], had an increase in amyloidosis compared to WT mice. In addition, when [Tlr(Lps-d)/Tlr(Lps-d)] mouse-derived microglia were stimulated with TLR4 ligands, an increase in Aβ uptake was not observed. However, this TLR4-mediated signaling apparently does not initiate at very early stages of pathology, since the same authors found that in TLR4-mutated mice Aβ production and Aβ deposition were not different from wild-type APP/PS1 mice at 5 months of age. Only at 9 months, Aβ levels increased, likely because the TLR4 mutation also diminished fibrillary Aβ-induced CCL3 expression in monocytes, suggesting the involvement of TLR4 signaling in the recruitment of microglia/monocytes. This was associated with an anticipation of cognitive deficits normally appearing around the age of 12 months in the APP/PS1 mice [34, 35].

The harmful effects of TLR4 in the context of AD raised instead from data demonstrating that TLR4-loss of function polymorphism protects against AD [36]. In a Drosophila model of AD expressing $A\beta_{1-42}$, TLR4 signaling elicited neurotoxic inflammation [37]. Also, the immune response, assessed through the quantification of TNFα release, induced by fibrillary Aβ when applied to human monocytic THP-1 cell line, was blocked by the concomitant application of antibodies against TLR2 and TLR4 [38]. An increased expression level of *tlr4* mRNA, accompanied by neurodegeneration mediated by an increased activity of C-Jun N-terminal kinase (JNK) and caspase 3, was detectable in vitro when mouse neuronal cultures were exposed to Aβ. Of note, TLR4 selective function elimination hindered these effects [39].

A *tlr4* mRNA–marked upregulation was described also in vivo in the APP23 transgenic mice, which carry a double human Swedish mutation on the APP gene [40, 41]. In this study, the authors micro-dissected plaques, tissue surrounding plaques and plaque-free brain tissue. TLRs upregulation was detected only in plaque material specifically for *tlr2, tlr4, tlr5, tlr7*, and *tlr9* mRNA. An increase in TLR4 protein levels was also described in the APP/PS1 AD mouse model together with an increment of pro-inflammatory cytokines, which was abolished in AD mice carrying a destructive mutation on the *tlr4* gene [41, 42].

Tlr4 gene upregulation has been observed also in the brains of AD patients, mainly in the temporal cortex region but also in the cerebellum [43–45]. However, this aspect remains controversial, since levels of TLR4 were found to be slightly decreased in tissue specimen of end-stage AD patients [39], although this was explained as a possible loss of TLR4-expressing neurons, because of TLR4-mediated predisposition of neurons to death [43].

In vivo and in vitro studies have demonstrated that binding of Aβ to TLR4 entails microglial activation and aberrant release of inflammatory mediators containing IL-1β, TNF-α, and reactive oxygen species (ROS) [46], leading to neuronal degeneration and accelerating the progression of AD [47]. In this context, TLR4 implication was also described for the high-mobility group box 1 (HMGB1) protein-mediated neurite degeneration. The authors demonstrated that HMGB1, normally released by necrotic or hyperexcited neurons, triggers hyperphosphorylation of myristoylated alanine-rich C-kinase substrate (MARCKS) at Ser46 then fostering neurite degeneration because of actin destabilization. MARCKS is a submembrane protein crucial to stabilize the actin network, which phosphorylation occurs at Ser46 prior to aggregation of Aβ, and is sustained throughout AD progression in both human and mouse brains. Of note, the authors demonstrated that TLR4-MAPK activation is required for the HMGB1-mediated MARCKS hyperphosphorylation. Indeed, TLR4 antagonist or knockdown by TLR4-shRNA suppressed it, whereas TLR4 ligand promoted this event [48].

A spontaneous loss-of-function mutation in the *Tlr4* gene strongly inhibited microglial and monocytic activation upon Aβ conditioning, resulting in significantly less release of the inflammatory cytokines IL-6, TNFα, and nitric oxide.

In addition, treatment of primary murine neuronal cells with media derived from Aβ-treated microglial cells indicated that TLR4 contributes to Aβ peptide–induced microglial neurotoxicity [43].

We also contributed to demonstrate the involvement of TLR4 in the context of AD by exploiting our Aβ oligomer–induced acute mouse model [19]. In this model, C57BL/6 naïve mice receiving one injection of a well-characterized solution of Aβ oligomers in the cerebral ventricle show a significant memory impairment in the novel object recognition test, which is associated with glial cell activation and an increase in pro-inflammatory cytokines such as IL-1β, IL-6, and TNF-α. However, if mice are pretreated with a cyanobacterial LPS TLR4 antagonist, their memory is preserved, indicating that Aβ oligomer-mediated memory impairment relies on the activation of TLR4. This finding was further confirmed by demonstrating that if AβOs were injected in the cerebroventricle of TLR4 knockout mice their effects on both memory and glial cells were hindered [20]. Other approaches in rats have demonstrated that intracerebral injection of Aβ induce cognitive deficits and neuroinflammation implicating an overexpression of TLR4 and increase in pro-inflammatory cytokines [22, 49].

Worth of note is the new evidence highlighting the fact that the Asp299Gly coding variant rs4986790 in the extracellular domain of TLR4 is associated with longevity [50] and a lower risk to develop AD [36]. The main hypothesis states that this TLR4 missense mutation attenuates the TLR4-mediated signaling, thus reducing neuroinflammation and the induction of neurodegenerative phenomena [51]. In this regard, Miron et al. [52], in the attempt to decipher the biological mechanism behind this protection, demonstrated that pre-symptomatic subjects with a parental history of AD carrying the rs4986790 (G) polymorphism did not display higher level of *tlr4* mRNA or protein, rather stable level of IL-1β in the CSF over time. In the same subjects, higher cortical thickness in areas of the brain involved in executive functions were observed, together with better performances in visuospatial and constructional tests. It is to be pinpointed that visuospatial perception has recently emerged as a potential biomarker of early AD detection [53]. In contrast, homozygous carriers for the major allele (A) exhibit a progressive increase in IL-1β throughout time and lower visuospatial abilities [52]. Based on this evidence, the authors hypothesized that this association TLR4 rs4986790 G variant-IL-1β CSF level may prevent AD by preserving cortical structure and those cognitive functions relying on the activity of these specific cortical brain areas.

3 TLR4 in PD

The contribution of TLR4 in α-synucleinopathies is controversial and debated as in the context of AD, although also in this case the harmful effects apparently predominate over the beneficial (Kouli et al., 2019). Indeed, despite human evidence highlighting an increased expression of TLR4 in the most affected brain areas, such as the substantia nigra [54] and the caudate putamen [55] in PD patients, in vivo and in vitro studies describe controversial findings on TLR4 contribution in the pathology.

The protective involvement of TLR4 emerged from studies showing that TLR4 ablation hampered the ability of microglia to phagocytose α-syn in vitro. In the same study, TLR4 ablation in PD mice overexpressing α-syn promoted an enhancement of motor disability and an increment of pro-inflammatory cytokines and dopaminergic neuronal cell death [56].

In contrast, the harmful role of TLR4 raised from experiments in different in vitro and in vivo models of PD. 1-Methyl-4-phenyl-1,2,3,6-tetrahydropyridine (MPTP) is a prodrug to the neurotoxin MPP+ which is used for the induction of typical PD symptoms and dopaminergic neuronal loss in in vivo models [57]. Using in vitro approaches, Zhou and colleagues demonstrated that TLR4 silencing hinders microglial and NF-κB activation induced by MPTP cell exposure, leading to a lower release of TNF-α, IL-1β, and iNOS [58]. Data from MPTP-treated mice revealed an upregulation of TLR4, as well as a reduced vulnerability to MPTP-induced pathological outcomes, such as motor disability, microglial and inflammasome activation, and dopaminergic neuronal loss, in the absence of TLR4 (siRNA, gene knockout) [59–62].

Several works also pointed out the involvement of TLR4 in mediating an increase in pro-inflammatory cytokine release from microglia and astrocytes when exposed to α-syn. The exposure of both $TLR4^{+/+}$ microglial cells and astrocytes to different moieties of wild type α-syn (soluble, full-length oligomers and C-terminally truncated) triggers their activation and the release of pro-inflammatory mediators such as TNF-α and IL-6. Remarkably, ablation of the receptor results in the suppression of the pro-inflammatory response in both cell types [63, 64]. Consistently, the involvement of TLR4 in mediating the astrocyte α-syn-induced pro-inflammatory response has also been reported by Rannikko and collaborators. In their study the authors described an α-syn dose-dependent increase in the transcriptional levels of different pro-inflammatory mediators such as IL-1β, IL-6, TNF-α, and COX-2 in $TLR4^{+/+}$, but not in $TLR4^{-/-}$ astrocytes [65]. This evidence was confirmed in a more recent study, where glial cells undergoing a 5-day exposure to picomolar concentration of α-syn oligomers, monomers, or fibrils were sensitized toward a pro-inflammatory response documented by a significant increase of TNF-α only upon oligomer exposure. No immune reaction was detectable when monomer or fibrils were applied to cells. The confirmation that the oligomer-mediated response was mediated by TLR4 was proved by the fact that TLR4 antagonist abolished it [66].

In contrast to these observations proving the involvement of TLR4 in the context of PD, we recently demonstrated in an α-syn oligomer-induced acute mouse model that TLR4 is not implied in mediating the α-syn oligomer-mediated memory impairment. Worth of note is the fact that in this mouse model, the ICV injection of α-syn oligomers induced a transient memory impairment associated with glial cell activation and an increase in pro-inflammatory cytokine expression. Anti-inflammatory drugs administered before α-syn oligomers fully prevented the memory impairment. However, in contrast to what we described above with Aβ oligomers, when α-syn oligomers were injected in the brain ventricle of TLR4 knockout mice, their memory was still impaired. Interestingly, this outcome was not replicated when the α-syn oligomers were ICV injected in mice pretreated with a TLR2 antagonist, which preserved mouse memory [21], thus ruling out the role of TLR4, at least, in the α-syn oligomer-mediated memory impairment.

Another interesting emerging aspect in the field of neurodegenerative diseases is the new relation between gut microbiota dysbiosis and an increased risk of developing AD and PD. In the latter case, a series of evidence highlight the fact that PD could much likely originate in the intestinal mucosa [67]. Alterations in gut bacterial composition have been found in PD patients as well as the presence of α-syn aggregates [67]. Transplantation of PD patients gut microbiota in germ-free mice overexpressing α-syn turns into an aggravated PD-related phenotype development [68]. Several evidences indicate that TLR4 are involved also in this context, since increase in *tlr4* mRNA levels were found in the intestinal mucosa biopsy of PD subjects, concomitantly to an increase in the expression of pro-inflammatory cytokines [69]. In addition, the same authors demonstrated in vivo, through the rotenone-induced mouse models which displays typical PD-related symptoms, that in the absence of TLR4 gut, motor, and brain abnormalities were reduced compared to wild-type rotenone-treated mice [69].

4 Targeting TLR4 in AD and PD: How and When

Based on the evidence described above, in terms of therapeutic intervention it appears that an inhibition of TLR4 more likely will provide the most appropriate therapeutic response.

A due consideration, however, is that glial cell response has a double-edge sword action being protective while exerting a reversible action of defense, and detrimental when its resolution is no longer achieved, thus perpetuating toward a chronic and neurotoxic state.

For these main reasons, it appears more conceivable to manage glial cell activation or inhibition in chronic pathologies such as AD and PD, by considering the different stages of pathology.

Indeed, it is indisputable that TLR4-dependent microglia-mediated clearance of misfolded proteins is a fundamental process which must be guaranteed. Apparently, Aβ and α-syn aggregates bind TLR4 and other receptors promoting their clearance in the early stages of disease, but foster neurotoxic glial cell activation and an impaired glial cell activity in the later stages [70, 71]. In line with this, it is well documented that an acute or chronic pro-inflammatory stimulus with LPS in APP/PS1 mice turns either into an enhanced Aβ phagocytosis or deposition, respectively. Indeed, while an acute treatment promotes Aβ clearance, chronic exposure favors a sustained neuroinflammatory process associated with a sustained Aβ production and a hampered ability of microglia to phagocytose it [72–75]. In contrast, if the pro-inflammatory process is induced through repeated injection of monophosphoryl lipid A, which binds TLR4 but induces only a moderate neuroimmune reaction, microglial phagocytosis is preserved, Aβ brain load is reduced, and memory recovery is observed [75]. Based on this evidence, it is crucial to take into consideration that TLR4 manipulation in a therapeutic prospective is not only a matter of how to manipulate it, but also a matter of when, with a TLR4 stimulation more conceivable in the early stages and a TLR4 inhibition advisable in the later stages of disease.

Some small molecules, natural compounds or repurposed drugs have been investigated for their efficacy in various neurodegenerative diseases. TAK242, or resatorvid, is a small molecule suppressing TLR4 activation. TAK242 has been tested preclinically in mouse models of traumatic brain injury and amyloid lateral sclerosis showing neuroprotective effects and improvement at behavioral level [76, 77]. TAK242 pretreatment also abolished α-syn-induced neuronal cell death and TNF-α release [66]. A short-term use (4 days) tolerability for TAK242 has been already proved in a phase 1 clinical trial for sepsis [78]. Vinpocetine is a repurposed drug, clinically used for cerebrovascular diseases [79] with anti-inflammatory activities. It was shown to lower TLR2 and TLR4 expression in peripheral monocytes of PD patients randomized in a double-blind placebo-controlled trial, and downregulate MyD88, NF-κB and TNF-α [80]. Candesartan cilexetil, which is licensed for the treatment of hypertension [81], reduced TLR4 expression and release of pro-inflammatory cytokines both in vitro and in vivo upon LPS or α-syn application [82].

Among the natural compounds, the most widely studied is curcumin, which was found to directly repress in vitro the TLR4-mediated pathway [83]. In MPTP-induced in vitro PD models, for instance, curcumin reduced astrocyte activation, the production of pro-inflammatory cytokines, as well as the activation of TLR4 signaling [84]. However, data are still not clear and further work is required before its application in the clinic.

5 Conclusion

Although the contribution of inflammation in the pathogenesis of neurodegenerative disorders, and in particular in AD and PD, has been strongly supported by genetic, epidemiological, and experimental studies, the translation of this evidence in the therapeutic approaches remains difficult and unsuccessful [85]. The limited efficacy of anti-inflammatory drugs to influence disease progression of AD and PD might have various explanations. In this context, the recent negative experiences suggest, together with adequate timing of treatment, to improve therapeutic approaches in several directions. The possibility that a single treatment can halt or attenuate the complex pathological scenario of neurodegenerative disorder, like AD or PD, is remote, therefore more realistic approaches should take in consideration the combination of drugs. Thus, the treatment with anti-inflammatory drugs could be associated with other treatment active on amyloid deposits and/or with neuroprotective activity. A second aspect to consider is the selection of the patients, and the treatment with anti-inflammatory drugs should be associated with a specific profile ascertained in the early phase of the disease. According to the principle of precision medicine, genetic and biological determinations might contribute to identify subjects more sensitive to anti-inflammatory treatment [86]. Finally, the studies summarized in this review suggest a more specific approach to the reactivity of immune system. Modulation of specific pharmacological target like TLR4, rather than classical anti-inflammatory compounds, might affect a limited population of AD or PD subjects but with more elevated chances to success.

References

1. Lippa CF, Saunders AM, Smith TW, Swearer JM, Drachman DA, Ghetti B, Nee L, Pulaski-Salo D, Dickson D, Robitaille Y, Bergeron C, Crain B, Benson MD, Farlow M, Hyman BT, George-Hyslop SP, Roses AD, Pollen DA. Familial and sporadic Alzheimer's disease: neuropathology cannot exclude a final common pathway. Neurology. 1996;46:406–12. https://doi.org/10.1212/wnl.46.2.406.
2. Papapetropoulos S, Adi N, Ellul J, Argyriou AA, Chroni E. A prospective study of familial versus sporadic Parkinson's disease. Neurodegener Dis. 2007;4:424–7. https://doi.org/10.1159/000107702.
3. Armstrong RA. Spatial patterns of β-amyloid (Aβ) deposits in familial and sporadic Alzheimer's disease. Folia Neuropathol. 2011;49:153–61.
4. Long JM, Holtzman DM. Alzheimer disease: an update on pathobiology and treatment strategies. Cell. 2019;179:312–39. https://doi.org/10.1016/j.cell.2019.09.001.
5. Simon DK, Tanner CM, Brundin P. Parkinson disease epidemiology, pathology, genetics, and pathophysiology. Clin Geriatr Med. 2020;36:1–12. https://doi.org/10.1016/j.cger.2019.08.002.
6. Soto C, Pritzkow S. Protein misfolding, aggregation, and conformational strains in neurodegenerative diseases. Nat Neurosci. 2018;21:1332–40. https://doi.org/10.1038/s41593-018-0235-9.
7. Baba M, Nakajo S, Tu PH, Tomita T, Nakaya K, Lee VM, Trojanowski JQ, Iwatsubo T. Aggregation of alpha-synuclein in Lewy bodies of sporadic Parkinson's disease and dementia with Lewy bodies. Am J Pathol. 1998;152:879–84.

8. Spillantini MG, Crowther RA, Jakes R, Hasegawa M, Goedert M. Alpha-Synuclein in filamentous inclusions of Lewy bodies from Parkinson's disease and dementia with Lewy bodies. Proc Natl Acad Sci USA. 1998;95:6469–73.
9. Borghi R, Marchese R, Negro A, Marinelli L, Forloni G, Zaccheo D, Abbruzzese G, Tabaton M. Full length alpha-synuclein is present in cerebrospinal fluid from Parkinson's disease and normal subjects. Neurosci Lett. 2000;287:65–7. S0304-3940(00)01153-8 [pii]
10. El-Agnaf OM, Salem SA, Paleologou KE, Cooper LJ, Fullwood NJ, Gibson MJ, Curran MD, Court JA, Mann DM, Ikeda S, Cookson MR, Hardy J, Allsop D. Alpha-synuclein implicated in Parkinson's disease is present in extracellular biological fluids, including human plasma. FASEB J. 2003;17:1945–7. https://doi.org/10.1096/fj.03-0098fje.
11. Ohrfelt A, Grognet P, Andreasen N, Wallin A, Vanmechelen E, Blennow K, Zetterberg H. Cerebrospinal fluid alpha-synuclein in neurodegenerative disorders-a marker of synapse loss? Neurosci Lett. 2009;450:332–5. https://doi.org/10.1016/j.neulet.2008.11.015.
12. Benilova I, Karran E, De Strooper B. The toxic Abeta oligomer and Alzheimer's disease: an emperor in need of clothes. Nat Neurosci. 15:349–57. https://doi.org/10.1038/nn.3028.
13. Mucke L, Selkoe DJ. Neurotoxicity of amyloid beta-protein: synaptic and network dysfunction. Cold Spring Harb Perspect Med. 2012;2:a006338. https://doi.org/10.1101/cshperspect.a006338.
14. Forloni G, Artuso V, La Vitola P, Balducci C. Oligomeropathies and pathogenesis of Alzheimer and Parkinson's diseases. Mov Disord. 2016;31:771–81. https://doi.org/10.1002/mds.26624.
15. Terry RD, Masliah E, Salmon DP, Butters N, DeTeresa R, Hill R, Hansen LA, Katzman R. Physical basis of cognitive alterations in Alzheimer's disease: synapse loss is the major correlate of cognitive impairment. Ann Neurol. 1991;30:572–80. https://doi.org/10.1002/ana.410300410.
16. Kuo YM, Emmerling MR, Vigo-Pelfrey C, Kasunic TC, Kirkpatrick JB, Murdoch GH, Ball MJ, Roher AE. Water-soluble Abeta (N-40, N-42) oligomers in normal and Alzheimer disease brains. J Biol Chem. 1996;271:4077–81.
17. McLean CA, Cherny RA, Fraser FW, Fuller SJ, Smith MJ, Beyreuther K, Bush AI, Masters CL. Soluble pool of Abeta amyloid as a determinant of severity of neurodegeneration in Alzheimer's disease. Ann Neurol. 1999;46:860–6.
18. Walsh DM, Klyubin I, Fadeeva JV, Cullen WK, Anwyl R, Wolfe MS, Rowan MJ, Selkoe DJ. Naturally secreted oligomers of amyloid beta protein potently inhibit hippocampal long-term potentiation in vivo. Nature. 2002;416:535–9. https://doi.org/10.1038/416535a.
19. Balducci C, Beeg M, Stravalaci M, Bastone A, Sclip A, Biasini E, Tapella L, Colombo L, Manzoni C, Borsello T, Chiesa R, Gobbi M, Salmona M, Forloni G. Synthetic amyloid-beta oligomers impair long-term memory independently of cellular prion protein. Proc Natl Acad Sci USA. 2010;107:2295–300. https://doi.org/10.1073/pnas.0911829107.
20. Balducci C, Frasca A, Zotti M, La Vitola P, Mhillaj E, Grigoli E, Iacobellis M, Grandi F, Messa M, Colombo L, Molteni M, Trabace L, Rossetti C, Salmona M, Forloni G. Toll-like receptor 4-dependent glial cell activation mediates the impairment in memory establishment induced by β-amyloid oligomers in an acute mouse model of Alzheimer's disease. Brain Behav Immun. 2017; https://doi.org/10.1016/j.bbi.2016.10.012.
21. La Vitola P, Balducci C, Cerovic M, Santamaria G, Brandi E, Grandi F, Caldinelli L, Colombo L, Morgese MG, Trabace L, Pollegioni L, Albani D, Forloni G. Alpha-synuclein oligomers impair memory through glial cell activation and via toll-like receptor 2. Brain Behav Immun. 2018;69:591–602. https://doi.org/10.1016/j.bbi.2018.02.012.
22. He Y, Zheng MM, Ma Y, Han XJ, Ma XQ, Qu CQ, Du YF. Soluble oligomers and fibrillar species of amyloid beta-peptide differentially affect cognitive functions and hippocampal inflammatory response. Biochem Biophys Res Commun. 2012;429:125–30. https://doi.org/10.1016/j.bbrc.2012.10.129.
23. Alvarez-Erviti L, Seow Y, Yin H, Betts C, Lakhal S, Wood MJ. Delivery of siRNA to the mouse brain by systemic injection of targeted exosomes. Nat Biotechnol. 2011;29:341–5. https://doi.org/10.1038/nbt.1807.

24. Couch Y, Alvarez-Erviti L, Sibson NR, Wood MJ, Anthony DC. The acute inflammatory response to intranigral alpha-synuclein differs significantly from intranigral lipopolysaccharide and is exacerbated by peripheral inflammation. J Neuroinflammation. 2011;8:166. https://doi.org/10.1186/1742-2094-8-166.
25. Klegeris A, Pelech S, Giasson BI, Maguire J, Zhang H, McGeer EG, McGeer PL. Alpha-synuclein activates stress signaling protein kinases in THP-1 cells and microglia. Neurobiol Aging. 2008;29:739–52. https://doi.org/10.1016/j.neurobiolaging.2006.11.013.
26. Heneka MT, Carson MJ, El Khoury J, Landreth GE, Brosseron F, Feinstein DL, Jacobs AH, Wyss-Coray T, Vitorica J, Ransohoff RM, Herrup K, Frautschy SA, Finsen B, Brown GC, Verkhratsky A, Yamanaka K, Koistinaho J, Latz E, Halle A, Petzold GC, Town T, Morgan D, Shinohara ML, Perry VH, Holmes C, Bazan NG, Brooks DJ, Hunot S, Joseph B, Deigendesch N, Garaschuk O, Boddeke E, Dinarello CA, Breitner JC, Cole GM, Golenbock DT, Kummer MP. Neuroinflammation in Alzheimer's disease. Lancet Neurol. 2015;14:388–405. https://doi.org/10.1016/S1474-4422(15)70016-5.
27. Balducci C, Forloni G. Novel targets in Alzheimer's disease: a special focus on microglia. Pharmacol Res. 2018;130:402–13. https://doi.org/10.1016/j.phrs.2018.01.017.
28. Lee Y, Lee S, Chang S-C, Lee J. Significant roles of neuroinflammation in Parkinson's disease: therapeutic targets for PD prevention. Arch Pharm Res. 2019;42:416–25. https://doi.org/10.1007/s12272-019-01133-0.
29. Alam Q, Alam MZ, Mushtaq G, Damanhouri GA, Rasool M, Kamal MA, Haque A. Inflammatory process in Alzheimer's and Parkinson's diseases: central role of cytokines. Curr Pharm Des. 2016;22:541–8. https://doi.org/10.2174/1381612822666151125000300.
30. Morris GP, Clark IA, Zinn R, Vissel B. Microglia: a new frontier for synaptic plasticity, learning and memory, and neurodegenerative disease research. Neurobiol Learn Mem. 2013;105:40–53. https://doi.org/10.1016/j.nlm.2013.07.002.
31. Kumar V. Toll-like receptors in the pathogenesis of neuroinflammation. J Neuroimmunol. 2019;332:16–30. https://doi.org/10.1016/j.jneuroim.2019.03.012.
32. McGeer PL, Rogers J, McGeer EG. Inflammation, anti-inflammatory agents and Alzheimer disease: the last 12 years. J Alzheimers Dis JAD. 2006;9:271–6. https://doi.org/10.3233/jad-2006-9s330.
33. von Bernhardi R, Tichauer JE, Eugenin J. Aging-dependent changes of microglial cells and their relevance for neurodegenerative disorders. J Neurochem. 2010;112:1099–114. https://doi.org/10.1111/j.1471-4159.2009.06537.x.
34. Tahara K, Kim HD, Jin JJ, Maxwell JA, Li L, Fukuchi K. Role of toll-like receptor signalling in Abeta uptake and clearance. Brain. 2006;129:3006–19. https://doi.org/10.1093/brain/awl249.
35. Song M, Jin J, Lim JE, Kou J, Pattanayak A, Rehman JA, Kim HD, Tahara K, Lalonde R, Fukuchi K. TLR4 mutation reduces microglial activation, increases Abeta deposits and exacerbates cognitive deficits in a mouse model of Alzheimer's disease. J Neuroinflammation. 2011;8:92. https://doi.org/10.1186/1742-2094-8-92.
36. Minoretti P, Gazzaruso C, Vito CD, Emanuele E, Bianchi M, Coen E, Reino M, Geroldi D. Effect of the functional toll-like receptor 4 Asp299Gly polymorphism on susceptibility to late-onset Alzheimer's disease. Neurosci Lett. 2006;391:147–9. https://doi.org/10.1016/j.neulet.2005.08.047.
37. Tan L, Schedl P, Song HJ, Garza D, Konsolaki M. The toll-->NFkappaB signaling pathway mediates the neuropathological effects of the human Alzheimer's Abeta42 polypeptide in Drosophila. PLoS One. 2008;3:e3966. https://doi.org/10.1371/journal.pone.0003966.
38. Udan ML, Ajit D, Crouse NR, Nichols MR. Toll-like receptors 2 and 4 mediate Abeta(1-42) activation of the innate immune response in a human monocytic cell line. J Neurochem. 2008;104:524–33. https://doi.org/10.1111/j.1471-4159.2007.05001.x.
39. Tang SC, Lathia JD, Selvaraj PK, Jo DG, Mughal MR, Cheng A, Siler DA, Markesbery WR, Arumugam TV, Mattson MP. Toll-like receptor-4 mediates neuronal apoptosis induced by amyloid beta-peptide and the membrane lipid peroxidation product 4-hydroxynonenal. Exp Neurol. 2008;213:114–21. https://doi.org/10.1016/j.expneurol.2008.05.014.

40. Frank S, Copanaki E, Burbach GJ, Muller UC, Deller T. Differential regulation of toll-like receptor mRNAs in amyloid plaque-associated brain tissue of aged APP23 transgenic mice. Neurosci Lett. 2009;453:41–4. https://doi.org/10.1016/j.neulet.2009.01.075.
41. Jin X, Liu M-Y, Zhang D-F, Zhong X, Du K, Qian P, Yao W-F, Gao H, Wei M-J. Baicalin mitigates cognitive impairment and protects neurons from microglia-mediated neuroinflammation via suppressing NLRP3 inflammasomes and TLR4/NF-κB signaling pathway. CNS Neurosci Ther. 2019;25:575–90. https://doi.org/10.1111/cns.13086.
42. Jin JJ, Kim HD, Maxwell JA, Li L, Fukuchi K. Toll-like receptor 4-dependent upregulation of cytokines in a transgenic mouse model of Alzheimer's disease. J Neuroinflammation. 2008;5:23. https://doi.org/10.1186/1742-2094-5-23.
43. Walter S, Letiembre M, Liu Y, Heine H, Penke B, Hao W, Bode B, Manietta N, Walter J, Schulz-Schuffer W, Fassbender K. Role of the toll-like receptor 4 in neuroinflammation in Alzheimer's disease. Cell Physiol Biochem. 2007;20:947–56. https://doi.org/10.1159/000110455.
44. Chakrabarty P, Li A, Ladd TB, Strickland MR, Koller EJ, Burgess JD, Funk CC, Cruz PE, Allen M, Yaroshenko M, Wang X, Younkin C, Reddy J, Lohrer B, Mehrke L, Moore BD, Liu X, Ceballos-Diaz C, Rosario AM, Medway C, Janus C, Li H-D, Dickson DW, Giasson BI, Price ND, Younkin SG, Ertekin-Taner N, Golde TE. TLR5 decoy receptor as a novel anti-amyloid therapeutic for Alzheimer's disease. J Exp Med. 2018;215:2247–64. https://doi.org/10.1084/jem.20180484.
45. Miron J, Picard C, Frappier J, Dea D, Théroux L, Poirier J. TLR4 gene expression and pro-inflammatory cytokines in Alzheimer's disease and in response to hippocampal Deafferentation in rodents. J Alzheimers Dis JAD. 2018;63:1547–56. https://doi.org/10.3233/JAD-171160.
46. Reed-Geaghan EG, Savage JC, Hise AG, Landreth GE. CD14 and toll-like receptors 2 and 4 are required for fibrillar a{beta}-stimulated microglial activation. J Neurosci. 2009;29:11982–92. https://doi.org/10.1523/JNEUROSCI.3158-09.2009.
47. Hanamsagar R, Hanke ML, Kielian T. Toll-like receptor (TLR) and inflammasome actions in the central nervous system. Trends Immunol. 2012;33:333–42. https://doi.org/10.1016/j.it.2012.03.001.
48. Fujita K, Motoki K, Tagawa K, Chen X, Hama H, Nakajima K, Homma H, Tamura T, Watanabe H, Katsuno M, Matsumi C, Kajikawa M, Saito T, Saido T, Sobue G, Miyawaki A, Okazawa H. HMGB1, a pathogenic molecule that induces neurite degeneration via TLR4-MARCKS, is a potential therapeutic target for Alzheimer's disease. Sci Rep. 2016;6:31895. https://doi.org/10.1038/srep31895.
49. Liu C-B, Wang R, Yi Y-F, Gao Z, Chen Y-Z. Lycopene mitigates β-amyloid induced inflammatory response and inhibits NF-κB signaling at the choroid plexus in early stages of Alzheimer's disease rats. J Nutr Biochem. 2018;53:66–71. https://doi.org/10.1016/j.jnutbio.2017.10.014.
50. Balistreri CR, Colonna-Romano G, Lio D, Candore G, Caruso C. TLR4 polymorphisms and ageing: implications for the pathophysiology of age-related diseases. J Clin Immunol. 2009;29:406–15. https://doi.org/10.1007/s10875-009-9297-5.
51. Figueroa L, Xiong Y, Song C, Piao W, Vogel SN, Medvedev AE. The Asp299Gly polymorphism alters TLR4 signaling by interfering with recruitment of MyD88 and TRIF. J Immunol Baltim Md. 2012;1950(188):4506–15. https://doi.org/10.4049/jimmunol.1200202.
52. Miron J, Picard C, Lafaille-Magnan M-É, Savard M, Labonté A, Breitner J, Rosa-Neto P, Auld D, Poirier J. PREVENT-AD research group, association of TLR4 with Alzheimer's disease risk and presymptomatic biomarkers of inflammation. Alzheimers Dement J Alzheimers Assoc. 2019;15:951–60. https://doi.org/10.1016/j.jalz.2019.03.012.
53. Mandal PK, Joshi J, Saharan S. Visuospatial perception: an emerging biomarker for Alzheimer's disease. J Alzheimers Dis JAD. 2012;31(Suppl 3):S117–35. https://doi.org/10.3233/JAD-2012-120901.
54. Shin W-H, Jeon M-T, Leem E, Won S-Y, Jeong KH, Park S-J, McLean C, Lee SJ, Jin BK, Jung UJ, Kim SR. Induction of microglial toll-like receptor 4 by prothrombin kringle-2: a potential pathogenic mechanism in Parkinson's disease. Sci Rep. 2015;5:14764. https://doi.org/10.1038/srep14764.

55. Drouin-Ouellet J, St-Amour I, Saint-Pierre M, Lamontagne-Proulx J, Kriz J, Barker RA, Cicchetti F. Toll-like receptor expression in the blood and brain of patients and a mouse model of Parkinson's disease. Int J Neuropsychopharmacol. 2015;18 https://doi.org/10.1093/ijnp/pyu103.
56. Stefanova N, Fellner L, Reindl M, Masliah E, Poewe W, Wenning GK. Toll-like receptor 4 promotes α-synuclein clearance and survival of nigral dopaminergic neurons. Am J Pathol. 2011;179:954–63. https://doi.org/10.1016/j.ajpath.2011.04.013.
57. Konnova EA, Swanberg M. Animal models of Parkinson's disease. In: Stoker TB, Greenland JC, editors. Parkinson's disease: pathogenesis and clinical aspects. Brisbane, AU: Codon Publications; 2018. http://www.ncbi.nlm.nih.gov/books/NBK536725/. Accessed 13 Dec 2019.
58. Zhou P, Weng R, Chen Z, Wang R, Zou J, Liu X, Liao J, Wang Y, Xia Y, Wang Q. TLR4 signaling in MPP+-induced activation of BV-2 cells. Neural Plast. 2016;2016:5076740. https://doi.org/10.1155/2016/5076740.
59. Noelker C, Morel L, Lescot T, Osterloh A, Alvarez-Fischer D, Breloer M, Henze C, Depboylu C, Skrzydelski D, Michel PP, Dodel RC, Lu L, Hirsch EC, Hunot S, Hartmann A. Toll like receptor 4 mediates cell death in a mouse MPTP model of Parkinson disease. Sci Rep. 2013;3:1393. https://doi.org/10.1038/srep01393.
60. Zhao X-D, Wang F-X, Cao W-F, Zhang Y-H, Li Y. TLR4 signaling mediates AP-1 activation in an MPTP-induced mouse model of Parkinson's disease. Int Immunopharmacol. 2016;32:96–102. https://doi.org/10.1016/j.intimp.2016.01.010.
61. Conte C, Roscini L, Sardella R, Mariucci G, Scorzoni S, Beccari T, Corte L. Toll like receptor 4 affects the cerebral biochemical changes induced by MPTP treatment. Neurochem Res. 2017;42:493–500. https://doi.org/10.1007/s11064-016-2095-6.
62. Campolo M, Paterniti I, Siracusa R, Filippone A, Esposito E, Cuzzocrea S. TLR4 absence reduces neuroinflammation and inflammasome activation in Parkinson's diseases in vivo model. Brain Behav Immun. 2019;76:236–47. https://doi.org/10.1016/j.bbi.2018.12.003.
63. Fellner L, Stefanova N. The role of glia in alpha-synucleinopathies. Mol Neurobiol. 2012;47:575–86. https://doi.org/10.1007/s12035-012-8340-3.
64. Fellner L, Jellinger KA, Wenning GK, Stefanova N. Glial dysfunction in the pathogenesis of alpha-synucleinopathies: emerging concepts. Acta Neuropathol. 2011;121:675–93. https://doi.org/10.1007/s00401-011-0833-z.
65. Rannikko EH, Weber SS, Kahle PJ. Exogenous α-synuclein induces toll-like receptor 4 dependent inflammatory responses in astrocytes. BMC Neurosci. 2015;16:57. https://doi.org/10.1186/s12868-015-0192-0.
66. Hughes CD, Choi ML, Ryten M, Hopkins L, Drews A, Botía JA, Iljina M, Rodrigues M, Gagliano SA, Gandhi S, Bryant C, Klenerman D. Picomolar concentrations of oligomeric alpha-synuclein sensitizes TLR4 to play an initiating role in Parkinson's disease pathogenesis. Acta Neuropathol (Berl). 2019;137:103–20. https://doi.org/10.1007/s00401-018-1907-y.
67. Yang D, Zhao D, Ali Shah SZ, Wu W, Lai M, Zhang X, Li J, Guan Z, Zhao H, Li W, Gao H, Zhou X, Yang L. The role of the gut microbiota in the pathogenesis of Parkinson's disease. Front Neurol. 2019;10:1155. https://doi.org/10.3389/fneur.2019.01155.
68. Sampson TR, Debelius JW, Thron T, Janssen S, Shastri GG, Ilhan ZE, Challis C, Schretter CE, Rocha S, Gradinaru V, Chesselet M-F, Keshavarzian A, Shannon KM, Krajmalnik-Brown R, Wittung-Stafshede P, Knight R, Mazmanian SK. Gut microbiota regulate motor deficits and neuroinflammation in a model of Parkinson's disease. Cell. 2016;167:1469–1480.e12. https://doi.org/10.1016/j.cell.2016.11.018.
69. Perez-Pardo P, Dodiya HB, Engen PA, Forsyth CB, Huschens AM, Shaikh M, Voigt RM, Naqib A, Green SJ, Kordower JH, Shannon KM, Garssen J, Kraneveld AD, Keshavarzian A. Role of TLR4 in the gut-brain axis in Parkinson's disease: a translational study from men to mice. Gut. 2019;68:829–43. https://doi.org/10.1136/gutjnl-2018-316844.
70. Fellner L, Irschick R, Schanda K, Reindl M, Klimaschewski L, Poewe W, Wenning GK, Stefanova N. Toll-like receptor 4 is required for α-synuclein dependent activation of microglia and astroglia. Glia. 2013;61:349–60. https://doi.org/10.1002/glia.22437.

71. Gambuzza ME, Sofo V, Salmeri FM, Soraci L, Marino S, Bramanti P. Toll-like receptors in Alzheimer's disease: a therapeutic perspective. CNS Neurol Disord Drug Targets. 2014;13:1542–58. https://doi.org/10.2174/1871527313666140806124850.
72. Sheng JG, Bora SH, Xu G, Borchelt DR, Price DL, Koliatsos VE. Lipopolysaccharide-induced-neuroinflammation increases intracellular accumulation of amyloid precursor protein and amyloid beta peptide in APPswe transgenic mice. Neurobiol Dis. 2003;14:133–45. https://doi.org/10.1016/s0969-9961(03)00069-x.
73. Lee JW, Lee YK, Yuk DY, Choi DY, Ban SB, Oh KW, Hong JT. Neuro-inflammation induced by lipopolysaccharide causes cognitive impairment through enhancement of beta-amyloid generation. J Neuroinflammation. 2008;5:37. https://doi.org/10.1186/1742-2094-5-37.
74. Blasko I, Marx F, Steiner E, Hartmann T, Grubeck-Loebenstein B. TNFalpha plus IFNgamma induce the production of Alzheimer beta-amyloid peptides and decrease the secretion of APPs. FASEB J. 1999;13:63–8.
75. Michaud JP, Halle M, Lampron A, Theriault P, Prefontaine P, Filali M, Tribout-Jover P, Lanteigne AM, Jodoin R, Cluff C, Brichard V, Palmantier R, Pilorget A, Larocque D, Rivest S. Toll-like receptor 4 stimulation with the detoxified ligand monophosphoryl lipid a improves Alzheimer's disease-related pathology. Proc Natl Acad Sci USA. 2013;110:1941–6. https://doi.org/10.1073/pnas.1215165110.
76. Fellner A, Barhum Y, Angel A, Perets N, Steiner I, Offen D, Lev N. Toll-like receptor-4 inhibitor TAK-242 attenuates motor dysfunction and spinal cord pathology in an amyotrophic lateral sclerosis mouse model. Int J Mol Sci. 2017;18 https://doi.org/10.3390/ijms18081666.
77. Feng Y, Gao J, Cui Y, Li M, Li R, Cui C, Cui J. Neuroprotective effects of Resatorvid against traumatic brain injury in rat: involvement of neuronal autophagy and TLR4 signaling pathway. Cell Mol Neurobiol. 2017;37:155–68. https://doi.org/10.1007/s10571-016-0356-1.
78. Rice TW, Wheeler AP, Bernard GR, Vincent J-L, Angus DC, Aikawa N, Demeyer I, Sainati S, Amlot N, Cao C, Ii M, Matsuda H, Mouri K, Cohen J. A randomized, double-blind, placebo-controlled trial of TAK-242 for the treatment of severe sepsis. Crit Care Med. 2010;38:1685–94. https://doi.org/10.1097/CCM.0b013e3181e7c5c9.
79. Zhang L, Yang L. Anti-inflammatory effects of vinpocetine in atherosclerosis and ischemic stroke: a review of the literature. Mol Basel Switz. 2014;20:335–47. https://doi.org/10.3390/molecules20010335.
80. Ping Z, Xiaomu W, Xufang X, Liang S. Vinpocetine regulates levels of circulating TLRs in Parkinson's disease patients. Neurol Sci Off J Ital Neurol Soc Ital Soc Clin Neurophysiol. 2019;40:113–20. https://doi.org/10.1007/s10072-018-3592-y.
81. Dasu MR, Riosvelasco AC, Jialal I. Candesartan inhibits toll-like receptor expression and activity both in vitro and in vivo. Atherosclerosis. 2009;202:76–83. https://doi.org/10.1016/j.atherosclerosis.2008.04.010.
82. Daniele SG, Béraud D, Davenport C, Cheng K, Yin H, Maguire-Zeiss KA. Activation of MyD88-dependent TLR1/2 signaling by misfolded α-synuclein, a protein linked to neurodegenerative disorders. Sci Signal. 2015;8:ra45. https://doi.org/10.1126/scisignal.2005965.
83. Youn HS, Saitoh SI, Miyake K, Hwang DH. Inhibition of homodimerization of toll-like receptor 4 by curcumin. Biochem Pharmacol. 2006;72:62–9. https://doi.org/10.1016/j.bcp.2006.03.022.
84. Yu S, Wang X, He X, Wang Y, Gao S, Ren L, Shi Y. Curcumin exerts anti-inflammatory and antioxidative properties in 1-methyl-4-phenylpyridinium ion (MPP(+))-stimulated mesencephalic astrocytes by interference with TLR4 and downstream signaling pathway. Cell Stress Chaperones. 2016;21:697–705. https://doi.org/10.1007/s12192-016-0695-3.
85. Ozben T, Ozben S. Neuro-inflammation and anti-inflammatory treatment options for Alzheimer's disease. Clin Biochem. 2019;72:87–9.
86. Forloni G, Balducci C. Alzheimer's disease, oligomers, and inflammation. J Alzheimers Dis. 2018;62:1261–76.

TLR4-Mediated Neuroinflammation in Human Induced Pluripotent Stem Cells and Cerebral Organoids

Massimiliano De Paola

Abstract The discovery of a reprogramming method to induce pluripotency in human somatic cells marked a dramatic turning point in the recent history of scientific progresses. Induced pluripotent stem cell (iPSCs) technology has been applied in regenerative and transplant medicine, disease modelling, drug screening, and studies on human developmental biology. The inflammatory responses elicited in healthy and diseased organisms have been extensively studied in animal models or artificial cell cultures, showing great usefulness but also some important limitations. Thus, the use of iPSCs as source of human cell cultures or organized tissues represented a great opportunity for neuroimmunology. In the latest years, the ability of iPSC-derived microglia to reproduce in vitro some of the genetic and phenotypic features of human adult microglia was shown. The development of platforms in which microglia, neurons, and macroglia derived from healthy or diseased subjects grow and mature in a single system allowed to demonstrate the role of toll-like receptor 4 (TLR4) in mediating neuroinflammation. The generation of three-dimensional (3D) microglia-embedded or -enriched organoids represent a dramatic further development of biologically relevant human-based models. In this context, the demonstrations of high similarity in gene expression and sensitivity to its endogenous and infective ligands in 3D constructs pose microglia-containing brain organoids in a prominent position for in-depth investigations on how TLR4 regulates immune cells interactions to orchestrate brain development and react to injury.

Keywords Human induced pluripotent stem cells · Brain organoids · Neuroinflammation · Immune cells · Spheroids · Microglia

M. De Paola (✉)
Department of Neuroscience, Mario Negri Institute for Pharmacological Research IRCCS, Milan, Italy
e-mail: massimiliano.depaola@marionegri.it

C. Rossetti, F. Peri (eds.), *The Role of Toll-Like Receptor 4 in Infectious and Non Infectious Inflammation*, Progress in Inflammation Research 87,
https://doi.org/10.1007/978-3-030-56319-6_8

1 Introduction

In 2007, Takahashi and Yamanaka reported a breakthrough method to reprogram and induce pluripotency in human adult differentiated cells by transduction of four basic transcription factors *Oct4*, *Sox2*, *Klf4*, and *c-Myc*, known as Yamanaka factors [25].

The reprogrammed cells were referred to as induced pluripotent stem cells (iPSCs) and paved the way for the development of new protocols for generating neural cells, among many other cell types, of human origin from an alternative source to fetal-derived tissues.

iPSCs are characterized by the properties of self-renewal and potency, since they are able to extensively proliferate and differentiate into specialized cell types derived from all the three primary germ layers: ectoderm, endoderm, or mesoderm [29].

Since that finding, it was possible to reprogram different types of mature cells, such as peripheral blood cells, fibroblasts, keratinocytes, and urine cells, to produce any types of cells, tissues, and organs, as for what was shown with embryonic stem cells. These features make iPSCs particularly suitable for regenerative and transplant medicine, disease modelling, drug screening, and studies on human developmental biology. Besides, a variety of protocols has been developed to induce ectoderm differentiation and subsequently proliferation, and differentiation and maturation of functionally active neuron, astrocyte, and oligodendrocyte, thus providing in vitro models for basic sciences.

By combining iPSC technique with specific neurodevelopmental factors able to induce cortical differentiation (i.e., embryoid-body like aggregation, [27]), three-dimensional (3D) brain organoids that resemble the structure, connection, and function of human central nervous system (CNS) were also generated in 2013 [11]. Since that year, different labs worldwide reported methods and protocols to obtain specialized region-specific brain organoids [10, 16, 18, 21–23]. Assembloids, composed by the fusion of different region-specific organoids, have been also generated in order to study brain neurodevelopmental events, such as neuronal migration and excitatory/inhibitory synapses interaction [4, 5].

Cerebral organoids are stem cell–derived models increasingly used in research studies to understand and interfere with human pathologies. They represent an exciting new technology intended to overcome the difficulties in accessing human neural tissues for research purpose.

These are 3D cell aggregates, usually derived from self-organizing pluripotent stem cells, which can have fetal origin or can be obtained from adult tissues by cell reprogramming. Recently, the organoid technology together with new powerful techniques for gene editing were brought to the forefront of biomedical research for their potential in investigating human disorders.

Cerebral organoids are particularly suitable for investigating different aspects of developmental processes and comparative biology, besides their flexibility and adaptability to model some features of brain diseases. Some aspects of these models are particularly valuable to promote valid translational results for human pathologies. For example, (1) no comparison between species is needed if organoids are derived from stem cells of human origin, (2) it is possible to obtain personalized organoids from patient-derived iPSCs, (3) toxicological studies on neural progenitors and different differentiation/maturation states can be performed, (4) genome-wide CRISPR-Cas9 screens can be applied.

One of the most interesting feature of brain organoids is the spontaneous development of different neural cell types, demonstrating cell–cell interactions with physiologically relevant 3D structures (e.g., myelinated axons showing electrophysiological properties in cortical spheroids; [19]). Most of the current protocols to generate organoids, however, consist of cells exclusively derived from the neuroectodermal lineage, thus limiting their applicability as they lack microglia. Mesodermal-derived microglia is, indeed, an important player of the immune-inflammatory mechanisms involved in neurodevelopment and diseases. It regulates normal brain development by guiding neuronal migration, strengthening neuronal connections, modulating oligodendrocyte and neuronal survival, besides protecting the brain against damage and infection. Microglia is also important in neurodegenerative diseases since its activation contributes to trigger, propagate, and sustain the detrimental neuroinflammatory events underlying large part of the pathological mechanisms of neurodegeneration. Currently, most of microglia models used to study neurodegenerative diseases come from murine tissues. These models are very useful for preclinical studies on the role of microglia either for direct analysis within the brain microenvironment (through organotypic slice cultures) or after explant from brain tissue of primary microglia cells, which can be maintained as purified cultures or used to establish cocultures with neuron and astrocyte for crosstalk investigations. However, they have some limitations, since translation of mouse results in human pathologies are usually very difficult, due to the differences in key modulators involved in neuroinflammatory events and the expression of risk genes [8].

To overcome this issue, scientists recently focused they efforts on developing new protocol for microglia derivation from human iPSC for single cell studies or enrichment of growing organoids, as described below.

2 Differentiation of Microglia from iPSC to Model Neuroinflammation and Neurodegenerative Diseases

Protocols for derivation and characterization of microglia from human iPSC have been only recently published. Most relied on pre-differentiation into hematopoietic stem cells (HSC) and myeloid progenitors by exposure to defined factors [9, 14, 20]. Besides the need of easy-to-use, reproducible, and efficient protocols for lab practitioners, in order to establish reliable models for the study of human microglia it is mandatory to obtain in vitro cultures composed of purified and functional microglia, which should reproduce the immune/inflammatory response, in addition to the closest genetic and transcriptional profile, as seen in vivo. Predictably, cocultures with other neural cells induce further cell maturation in terms of receptor expression, cytokine release and functional interaction to the human iPSC–derived microglia-like cells derived by these protocols. Indeed, cocultures of HSC on human astrocytes with subsequent differentiation protocol generate microglia with similar phenotype, gene expression profile, and functional properties of brain-isolated microglia [20]. The importance of neuron–microglia interaction for the microglia maturation was recently shown, even if results in this paper were mostly based on mouse iPSC microglia and only proof of principle data was shown for human iPSC microglia [26]. Microglial signature of iPSC-derived microglia can be significantly improved when cocultured with differentiating neuro-glial feeder layer and even more if added in neural spheroids [14]. The group of Dr. Cowley, from the Oxford University, showed that when human iPSC–derived microglia were cocultured with iPSC-derived cortical neurons, they express key microglia-specific markers and neurodegenerative disease–relevant genes, develop highly dynamic ramifications, and are phagocytic [9]. Importantly, cocultured microglia express relevant proteins for the mediation of inflammatory responses (i.e., CD11b, the LPS co-receptor CD14, CD45). Indeed, upon activation by LPS/interferon-gamma (IFNγ) they become reactive and switch to an ameboid phenotype releasing multiple microglia-relevant cytokines. In addition, microglia by this protocol showed downregulation of pathogen-response pathways, upregulation of the homeostatic functions, and promotion of a more anti-inflammatory cytokine response than corresponding monocultures, demonstrating that cocultures are preferable for modelling authentic microglia physiology [9]. Cells with similar features to cultured human adult and fetal microglia by both transcriptomic and functional analyses can be derived from iPSC by providing cues that mimic the environment present in the developing embryo [2]. Cultured microglia obtained through this two-step maturation protocol secrete cytokines in response to inflammatory stimuli, migrate and undergo calcium transients, and robustly phagocytose CNS substrates. Indeed, depending on their cell surface receptor stimuli (i.e., IFNγ, IL-1β, or LPS), iPSC-derived microglia differentially release cytokines/chemokines, a feature that closely resemble the responses observed in isolated primary microglia. In particular, LPS exposure induced a robust induction of 10 measured cytokines, including the classical pro-inflammatory cytokines TNFα, IL1α, IL-6 [2].

Besides the clear significance of these models for the study of mechanistic involvement of "healthy" microglia responses to external stimuli, even more relevant could be the possibility that they offer to reproduce "diseased" human microglia in vitro. Muffat et al., for example, published a robust protocol to generate and maintain microglia from multiple disease-specific cell lines and find that microglia derived from patients affected by Rett syndrome are smaller than their isogenic controls [14]. Functionally active microglia were also used to examine the effects of Aβ fibrils and brain-derived tau oligomers on AD-related gene expression and to interrogate mechanisms involved in synaptic pruning, thus providing evidence of disease-specific neurodegenerative mechanisms [2].

Addressing the role of TLR4 modulation in disease-specific human immune cells with such relevant models could open new prospective for disease pathogenesis and treatment with highly translational significance for human pathologies.

3 Microglia-Enriched 3D Brain Organoids to Study Neuroinflammation and Human Neurodegenerative Diseases

To make a step forward toward complete and reliable models for the study of neuroinflammation in human pathologies, new methods to obtain microglia-enriched iPSC-derived brain organoids have been recently reported.

Addition of microglia, either as immortalized cell lines [1] or derived from stem cells [14, 15, 24], to region-specific brain spheroids has proven to generate physiologically relevant 3D models for degenerative disease. Indeed, different pathologies have been recently modelled in vitro by these protocols, as for example, dengue and Zika virus infection [1, 15], Rett-syndrome caused by *MECP2* mutation [14], and Alzheimer's disease [13]. The stimulation with interferon (IFNγ) and endotoxin (LPS) induces microglia in single cultures to release chemokines and cytokines (CXCL10, CCL3, IL-6, and TNF-α, in particular) above baseline. However, when embedded in 3D spheroids, they assume a ramified, resting morphology and are able to switch to an amoeboid and actively migrating state in response to injury driven by (a) a mechanical damage of the spheroids by a needle [2]; or (b) focal laser injury, or ATP and ADP release from dying cells via purinergic receptors, such as P2RY12/13 [14]. Furthermore, when added to region-specific brain spheroids, microglia showed increased TLR4 gene expression and are sensitive to an NFκb inhibitor that was active in reducing inflammation mediated by Aβ42 oligomer stimulation [9].

Recently, a group from the Utrecht University found that microglia innately develop within iPSC-derived cerebral organoids generated with a modified protocol from the original work of Lancaster et al. [11]. Those microglia perform similarly to primary human microglia in functional assays, as for responses to inflammatory stimuli. After exposure to *E. Coli* LPS, indeed, mRNA expression of IL-6 and IL1β was significantly increased in single-cell suspensions prepared from fragmented organoids [17]. When the whole organoid (including microglia) was exposed to LPS, further increase in IL-6 and TNFα was measured after 24 and 72 hours, demonstrating the high sensitivity of immune cells in such construct.

4 Conclusions

TLR4 role in immune activation and neuroinflammation in the CNS has been extensively described. Apart from the key role in recognizing invading pathogens, different molecules produced or circulating during abnormal situations, such as during tissue damage, are able to trigger TLR4-dependent signaling [28]. Even if toll-like receptors showed high conservative structures across the species, evolutionary processes lead to substantial diversity in affinity and specificity to its ligands, the TLR4 gene, and cellular expression patterns and tissue distribution. Consequently, TLR4 functions vary across different species, and the results of receptor activation might substantially change among the organisms. For example, the TLR4 expression pattern in the mouse CNS differs from the human one since only microglial cells express TLR4, but not astrocytes nor oligodendrocytes [6]. In mouse CNS cell cultures, indeed, LPS induces significant injury to neurons [7] or developing oligodendrocytes [12], only in the presence of microglia. Given that TLR4-dependent mechanisms are largely investigated in preclinical animal models and that TLR4 ligands have shown increasing involvement in clinical applications (as for vaccine adjuvant [3]), the extent to which an animal model represents and predicts the human condition is of particular importance.

The difficulties in reaching tissue samples from human CNS warranted the development of new models and tools to unveil the cell-specific neuroinflammatory mechanisms in human diseases. In this sense, the birth and development of iPSC-derived human cell technologies represent a scientific revolution for this research area. The results obtained in the very latest years, as reported above, showed, in fact, that is now possible to establish genetically defined, highly reproducible cell lines from patients or isogenic control as valuable tool for the study of neuroinflammation. In particular, the development of new protocols to obtain functionally active microglia with well-defined genetic background brought these models to the forefront of biomedical research for their translational potential in neuroinflammatory and neurodegenerative disease modelling. The generation of 3D microglia-embedded or -enriched organoids represent a dramatic further development of biologically relevant human-based models. The development of platforms in which microglia, neurons, and macroglia grow and mature in a single system holds, indeed, the promise for a better understanding of how these cells interact to orchestrate brain development and react to injury. In this context, the demonstrations of increase gene expression and sensitivity to its endogenous and infective ligands, as described above, pose TLR4 in a prominent position for future investigations on the mechanisms underlying neuroinflammatory onset and propagation in health and diseases.

In conclusion, the evidence reported in this chapter supports the idea that the organoid-microglia model represents a valuable tool to understand the interactions of different cell types in the human brain and the role of microglia in human CNS inflammation.

References

1. Abreu CM, Gama L, Krasemann S, Chesnut M, Odwin-Dacosta S, Hogberg HT, Hartung T, Pamies D. Microglia increase inflammatory responses in iPSC-derived human brain spheres. Front Microbiol. 2018;9:2766. https://doi.org/10.3389/fmicb.2018.02766.
2. Abud EM, Ramirez RN, Martinez ES, Healy LM, Nguyen CHH, Newman SA, Yeromin AV, Scarfone VM, Marsh SE, Fimbres C, Caraway CA, Fote GM, Madany AM, Agrawal A, Kayed R, Gylys KH, Cahalan MD, Cummings BJ, Antel JP, Mortazavi A, Carson MJ, Poon WW, Blurton-Jones M. iPSC-derived human microglia-like cells to study neurological diseases. Neuron. 2017;94:278–293.e9. https://doi.org/10.1016/j.neuron.2017.03.042.
3. Alving CR, Peachman KK, Rao M, Reed SG. Adjuvants for human vaccines. Curr Opin Immunol. 2012;24:310–5. https://doi.org/10.1016/j.coi.2012.03.008.
4. Bagley JA, Reumann D, Bian S, Lévi-Strauss J, Knoblich JA. Fused cerebral organoids model interactions between brain regions. Nat Methods. 2017;14:743–51. https://doi.org/10.1038/nmeth.4304.
5. Birey F, Andersen J, Makinson CD, Islam S, Wei W, Huber N, Fan HC, Metzler KRC, Panagiotakos G, Thom N, O'Rourke NA, Steinmetz LM, Bernstein JA, Hallmayer J, Huguenard JR, Paşca SP. Assembly of functionally integrated human forebrain spheroids. Nature. 2017;545:54–9. https://doi.org/10.1038/nature22330.
6. Bsibsi M, Ravid R, Gveric D, van Noort JM. Broad expression of toll-like receptors in the human central nervous system. J Neuropathol Exp Neurol. 2002;61:1013–21. https://doi.org/10.1093/jnen/61.11.1013.
7. De Paola M, Mariani A, Bigini P, Peviani M, Ferrara G, Molteni M, Gemma S, Veglianese P, Castellaneta V, Boldrin V, Rossetti C, Chiabrando C, Forloni G, Mennini T, Fanelli R. Neuroprotective effects of Toll-like receptor 4 antagonism in spinal cord cultures and in a mouse model of motor neuron degeneration. Mol Med. 2012;18:971–81. https://doi.org/10.2119/molmed.2012.00020.
8. Haenseler W, Rajendran L. Concise review: modeling neurodegenerative diseases with human pluripotent stem cell-derived microglia. Stem Cells. 2019;37:724–30. https://doi.org/10.1002/stem.2995.
9. Haenseler W, Sansom SN, Buchrieser J, Newey SE, Moore CS, Nicholls FJ, Chintawar S, Schnell C, Antel JP, Allen ND, Cader MZ, Wade-Martins R, James WS, Cowley SA. A highly efficient human pluripotent stem cell microglia model displays a neuronal-co-culture-specific expression profile and inflammatory response. Stem Cell Rep. 2017;8:1727–42. https://doi.org/10.1016/j.stemcr.2017.05.017.
10. Hogberg HT, Bressler J, Christian KM, Harris G, Makri G, O'Driscoll C, Pamies D, Smirnova L, Wen Z, Hartung T. Toward a 3D model of human brain development for studying gene/environment interactions. Stem Cell Res Ther. 2013;4:S4. https://doi.org/10.1186/scrt365.
11. Lancaster MA, Renner M, Martin C-A, Wenzel D, Bicknell LS, Hurles ME, Homfray T, Penninger JM, Jackson AP, Knoblich JA. Cerebral organoids model human brain development and microcephaly. Nature. 2013;501:373–9. https://doi.org/10.1038/nature12517.
12. Lehnardt S, Lachance C, Patrizi S, Lefebvre S, Follett PL, Jensen FE, Rosenberg PA, Volpe JJ, Vartanian T. The Toll-like receptor TLR4 is necessary for lipopolysaccharide-induced oligodendrocyte injury in the CNS. J Neurosci. 2002;22:2478. https://doi.org/10.1523/JNEUROSCI.22-07-02478.2002.
13. Lin Y-T, Seo J, Gao F, Feldman HM, Wen H-L, Penney J, Cam HP, Gjoneska E, Raja WK, Cheng J, Rueda R, Kritskiy O, Abdurrob F, Peng Z, Milo B, Yu CJ, Elmsaouri S, Dey D, Ko T, Yankner BA, Tsai L-H. APOE4 causes widespread molecular and cellular alterations associated with Alzheimer's disease phenotypes in human iPSC-derived brain cell types. Neuron. 2018;98:1294. https://doi.org/10.1016/j.neuron.2018.06.011.

14. Muffat J, Li Y, Yuan B, Mitalipova M, Omer A, Corcoran S, Bakiasi G, Tsai L-H, Aubourg P, Ransohoff RM, Jaenisch R. Efficient derivation of microglia-like cells from human pluripotent stem cells. Nat Med. 2016;22:1358–67. https://doi.org/10.1038/nm.4189.
15. Muffat J, Li Y, Omer A, Durbin A, Bosch I, Bakiasi G, Richards E, Meyer A, Gehrke L, Jaenisch R. Human induced pluripotent stem cell-derived glial cells and neural progenitors display divergent responses to Zika and dengue infections. Proc Natl Acad Sci. 2018;115:7117. https://doi.org/10.1073/pnas.1719266115.
16. Muguruma K, Nishiyama A, Kawakami H, Hashimoto K, Sasai Y. Self-organization of polarized cerebellar tissue in 3D culture of human pluripotent stem cells. Cell Rep. 2015;10:537–50. https://doi.org/10.1016/j.celrep.2014.12.051.
17. Ormel PR, Vieira de Sá R, van Bodegraven EJ, Karst H, Harschnitz O, Sneeboer MAM, Johansen LE, van Dijk RE, Scheefhals N, Berdenis van Berlekom A, Ribes Martínez E, Kling S, MacGillavry HD, van den Berg LH, Kahn RS, Hol EM, de Witte LD, Pasterkamp RJ. Microglia innately develop within cerebral organoids. Nat Commun. 2018;9:4167. https://doi.org/10.1038/s41467-018-06684-2.
18. Pamies D, Hartung T, Hogberg HT. Biological and medical applications of a brain-on-a-chip. Exp Biol Med. 2014;239:1096–107. https://doi.org/10.1177/1535370214537738.
19. Pamies D, Barreras P, Block K, Makri G, Kumar A, Wiersma D, Smirnova L, Zhang C, Bressler J, Christian K, Harris G, Ming G-l, Berlinicke C, Kyro K, Song H, Pardo C, Hartung T, Hogberg H. A human brain microphysiological system derived from induced pluripotent stem cells to study neurological diseases and toxicity. ALTEX – Altern Anim Exp. 2017;34:362–76. https://doi.org/10.14573/altex.1609122.
20. Pandya H, Shen MJ, Ichikawa DM, Sedlock AB, Choi Y, Johnson KR, Kim G, Brown MA, Elkahloun AG, Maric D, Sweeney CL, Gossa S, Malech HL, McGavern DB, Park JK. Differentiation of human and murine induced pluripotent stem cells to microglia-like cells. Nat Neurosci. 2017;20:753–9. https://doi.org/10.1038/nn.4534.
21. Paşca AM, Sloan SA, Clarke LE, Tian Y, Makinson CD, Huber N, Kim CH, Park J-Y, O'Rourke NA, Nguyen KD, Smith SJ, Huguenard JR, Geschwind DH, Barres BA, Paşca SP. Functional cortical neurons and astrocytes from human pluripotent stem cells in 3D culture. Nat Methods. 2015;12:671–8. https://doi.org/10.1038/nmeth.3415.
22. Qian X, Nguyen HN, Song MM, Hadiono C, Ogden SC, Hammack C, Yao B, Hamersky GR, Jacob F, Zhong C, Yoon K, Jeang W, Lin L, Li Y, Thakor J, Berg DA, Zhang C, Kang E, Chickering M, Nauen D, Ho C-Y, Wen Z, Christian KM, Shi P-Y, Maher BJ, Wu H, Jin P, Tang H, Song H, Ming G. Brain-region-specific organoids using mini-bioreactors for modeling ZIKV exposure. Cell. 2016;165:1238–54. https://doi.org/10.1016/j.cell.2016.04.032.
23. Sakaguchi H, Kadoshima T, Soen M, Narii N, Ishida Y, Ohgushi M, Takahashi J, Eiraku M, Sasai Y. Generation of functional hippocampal neurons from self-organizing human embryonic stem cell-derived dorsomedial telencephalic tissue. Nat Commun. 2015;6:8896. https://doi.org/10.1038/ncomms9896.
24. Song L, Yuan X, Jones Z, Vied C, Miao Y, Marzano M, Hua T, Sang Q-XA, Guan J, Ma T, Zhou Y, Li Y. Functionalization of brain region-specific spheroids with isogenic microglia-like cells. Sci Rep. 2019;9:11055. https://doi.org/10.1038/s41598-019-47444-6.
25. Takahashi K, Tanabe K, Ohnuki M, Narita M, Ichisaka T, Tomoda K, Yamanaka S. Induction of pluripotent stem cells from adult human fibroblasts by defined factors. Cell. 2007;131:861–72. https://doi.org/10.1016/j.cell.2007.11.019.
26. Takata K, Kozaki T, Lee CZW, Thion MS, Otsuka M, Lim S, Utami KH, Fidan K, Park DS, Malleret B, Chakarov S, See P, Low D, Low G, Garcia-Miralles M, Zeng R, Zhang J, Goh CC, Gul A, Hubert S, Lee B, Chen J, Low I, Shadan NB, Lum J, Wei TS, Mok E, Kawanishi S, Kitamura Y, Larbi A, Poidinger M, Renia L, Ng LG, Wolf Y, Jung S, Önder T, Newell E, Huber T, Ashihara E, Garel S, Pouladi MA, Ginhoux F. Induced-pluripotent-stem-cell-derived primitive macrophages provide a platform for modeling tissue-resident macrophage differentiation and function. Immunity. 2017;47:183–198.e6. https://doi.org/10.1016/j.immuni.2017.06.017.

27. Watanabe K, Kamiya D, Nishiyama A, Katayama T, Nozaki S, Kawasaki H, Watanabe Y, Mizuseki K, Sasai Y. Directed differentiation of telencephalic precursors from embryonic stem cells. Nat Neurosci. 2005;8:288–96. https://doi.org/10.1038/nn1402.
28. Werling D, Jungi TW. TOLL-like receptors linking innate and adaptive immune response. Vet Immunol Immunopathol. 2003;91:1–12. https://doi.org/10.1016/S0165-2427(02)00228-3.
29. Wobus AM, Boheler KR. Embryonic stem cells: prospects for developmental biology and cell therapy. Physiol Rev. 2005;85:635–78. https://doi.org/10.1152/physrev.00054.2003.

The Role of TLR4 in Neural Stem Cells–Mediated Neurogenesis and Neuroinflammation

Lidia De Filippis and Francesco Peri

Abstract Stem cells in brain niches are responsible for neurogenesis and integration of new neurons into functional circuits, with inflammatory and immune system mediators playing critical roles in neurogenesis and in several diseases of the nervous system. TLR4 is known to be a master player in the development of neuroinflammation processes and to be upregulated in several animal models of neurodegenerative disorders like amyotrophic lateral sclerosis (ALS). Nonetheless, Toll-like receptor 4 (TLR4) also plays a key role in CNS homeostasis. TLR4 is expressed in microglia, which is a master player in neuroinflammatory processes, as well as in neural cells: astrocytes, neurons, oligodendrocytes, neural progenitors (NPCs), and neural stem cells (NSCs). Here we discuss the dual role of TLR4 in brain homeostasis, suggesting that, in a translational perspective, TLR4-mediated regulation of human neural stem cells (hNSCs) needs to be deeply investigated either for the identification of novel biomarkers for rare diseases or future therapeutic approaches of aging and neurodegenerative disorders.

Keywords Neural stem cells (NSC) · TLR4 · Neurogenesis · Neuroinflammation

1 Neural Stem Cells (NSC)

While during embryonic development pluripotent embryonic stem cells (ESC) are committed to rapidly generate the progenitors that will form the different tissues, on adulthood the presence of tissue-specific stem cells guarantees the tissue homeostasis and the maintenance of the stem compartment itself. The discovery of NSC in adult mammals marks a milestone in understanding the plasticity of the adult brain.

L. De Filippis (✉)
Dompé farmaceutici S.p.A, Milan, Italy
e-mail: lidia.defilippis@dompe.com

F. Peri
Department of Biotechnology and Biosciences, University of Milano Bicocca, Milan, Italy
e-mail: francesco.peri@unimib.it

C. Rossetti, F. Peri (eds.), *The Role of Toll-Like Receptor 4 in Infectious and Non Infectious Inflammation*, Progress in Inflammation Research 87,
https://doi.org/10.1007/978-3-030-56319-6_9

Neural stem cells are self-renewing, multipotent progenitors that reside in the nervous system [1].

Since specific and unequivocal NSC markers have still to be identified, the current concept of self-renewing and multipotent NSC in the adult mammalian brain has been largely based on in vitro studies demonstrating NSC stemness *a posteriori*. Cells capable of long-term expansion and differentiation into neurons and glia have been derived from adult rodent [2] and human brains [3], and cultured as neurospheres in a growth medium containing basic Fibroblast Growth Factor (bFGF) and Epidermal Growth Factor (EGF) (through the Neurosphere Assay) [1, 4]. In vitro studies have been then pivotal to the identification of bona fide NSC as inherently endowed with the following functional properties:

1. Self-renewal: The capacity to extensively proliferate and to auto-maintain.
2. Multipotency: The capacity to differentiate into the three neural lineages – astrocytes, neurons, and oligodendrocytes.
3. Plasticity: Flexibility to change both proliferation and differentiation potential according to environmental signaling and conditions.

In the adult CNS, stem cells are generally quiescent, but they are able to undergo activation and enter a cycle of continuous proliferation over long periods of time and to divide symmetrically or asymmetrically. This dynamic turnover is finely regulated by complex signaling in a temporal and spatial fashion [5].

NSC reside in specific areas called niches in adult CNS, defined as the microenvironment that intimately supports and tightly regulates stem cell behaviors, including their maintenance, self-renewal, fate specification, and development. The niche must preserve the self-maintenance of the stem cell compartment. It provides a trophic support, feeding NSCs through signals that regulate both their proliferation and their differentiation in a balanced manner. The cells of the niche, as well as retaining the stem cells in situ, retrieve them to the niche itself, a process called homing. This process is very important for the correct location of endogenous NSC during brain development but also in stem cell-mediated therapies, for addressing exogenous NSC to a stable and permissive environment after transplantation. A unique niche structure in the adult human brain has been recognized in two discrete regions, the ventricular-subventricular zone of the lateral ventricles (v-SVZ) and the subgranular zone (SGZ) in the dentate gyrus of the hippocampus [6, 7] (Fig. 1). Astroglia [6, 8], ependymal cells, vascular cells, NSC progeny, and mature neurons [9] are among major cellular components of the neurogenic niche.

Vascularization also plays a fundamental role in adult neurogenesis: The endothelial cells and the pericytes that surround the lumen of the vessels are separated from the brain parenchyma by the basal lamina which facilitates the activation of specific factors that allow regulating both neurogenesis, which takes place in tight places associated with blood vessels, and angiogenesis. In the recent years, the concept of "ectopic perivascular niche" has introduced the participation of the vascular blood system to the signaling that regulates also exogenous NSC after transplantation [10].

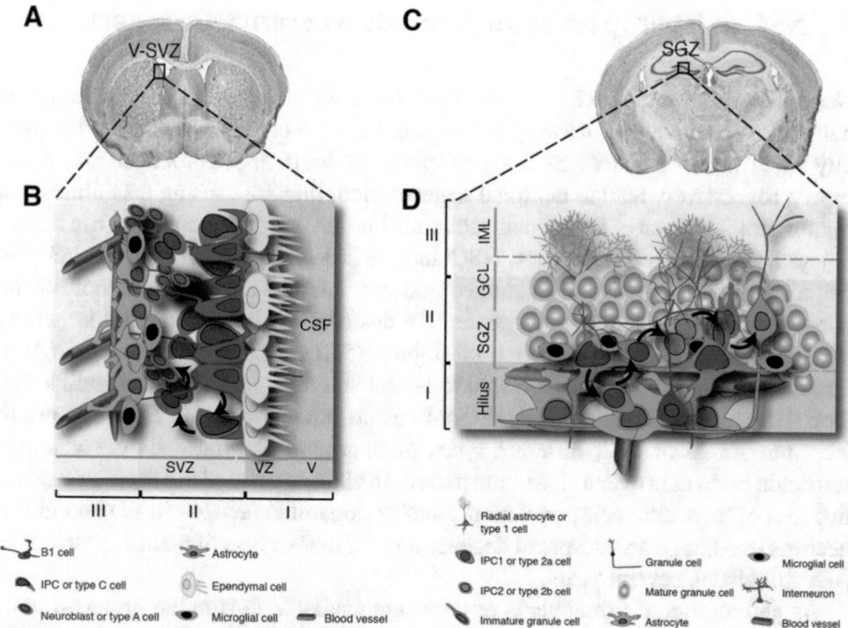

Fig. 1 *NSC and niche organization*: schematic representations of the adult V-SVZ and SGZ neurogenic compartments. (**a**) and (**c**), coronal sections of the adult mouse brain showing the localization of the V-SVZ and SGZ of the hippocampus. (**b**) and (**d**), cytoarchitecture of the V-SVZ (**b**), and of the SGZ of the DG of the hippocampus (**d**) in the adult mammalian brain. (**b**) Composition of the B1 cell domain into the V-SVZ. NSCs or type B1 cells (blue) extend from the proximal domain (domain I, dark grey) to the distal domain (domain III, light gray). At the level of the ventricles, B1 cells contact the CSF with their primary cilium extruding in the center of a rosette of multi-ciliated ependymal cells (yellow), forming the typical pinwheel-like structures on the ventricular surface. Here, NSCs can sense different signals circulating into the CSF. In the distal domain, type B1 cells contact the blood vessels (red) with their specialized end-foot terminations. In the intermediate domain (or domain II) type B1 cells give rise to IPCs (or type C cells, green), which are transit-amplifying cells generating neuroblasts (or type A cells, red). In this domain they are also in contact with their progeny, neighboring cells and neuronal terminations. D, composition of the RA domain at the level of the DG of the SGZ. RAs (or type 1 cells, blue) extend from the hilus of the hippocampus (domain I, dark gray) to the IML (distal domain or domain III, light gray). At the level of domain I, RAs sense the hilus microenvironment with their primary cilium and contact other RAs, IPCs, and blood vessels (red). RAs extend through their main shaft into the distal domain where their arborizations receive signals from glial cells and neuronal terminations. RAs give rise to IPCs that mature (trough blue IPC1 or type 2a cells, and light green IPC2 or type 2b cells) and differentiate into immature granule cells (IGC, red). During their maturation, IPCs move from the proximal domain to the intermediate domain (or domain II, composed by SGZ and GCL), where RAs receive signals from the progeny, neighboring NSCs, interneurons (purple), and microglia (gray). Finally, IGC differentiate into mature GC (green), which extend their axons into the hilus and arbores dendrites into the distal domain. Only few newborn neurons survive and become a long-lasting GC (pink). (From Ref. [11])

2 NSC as Therapeutics for Neurodegenerative Disorders

Neurodegenerative disease is a term used for a wide range of acute and chronic conditions in which neurons and glial cells in the CNS undergo damage and eventually are lost. Either under hereditary or sporadic conditions, neurodegenerative diseases progressively lead to neuronal degeneration, thereby causing disabilities and finally death, with an ensuing socioeconomic burden due to a decrease in life expectancy. Thanks to their properties, NSC and NSC-deriving progenitors (neural progenitor cell – NPC) have been studied and promoted as a therapeutic tool for the cure of many acute and neurodegenerative disorders including cerebral ischemic/hemorrhagic stroke and spinal cord injury (SCI), Alzheimer's disease (AD), Parkinson's disease (PD), amyotrophic lateral sclerosis (ALS), Huntington's disease (HD), and multiple sclerosis (MS). In acute cases, for example, in response to ischemic stroke or SCI, different types of neurons and glial cells die within a restricted brain area over a short time period. In chronic cases, there is either a selective loss of a specific cell population, such as dopamine neurons in PD and motor neurons in ALS, or a widespread degeneration of many types of neurons, such as in AD, still lasting several years.

As yet no effective treatments or cures are available to stop the progression of most neurodegenerative diseases. Pharmacological therapies are the only feasible approaches, with the only transient effect to alleviate and delay the progression of symptomatology of the disease. Stem cell–based therapy has been then promoted as a new perspective in translational medicine.

While ESCs were originally excluded from clinical application due to the etic issues related to their tissue source and to their tumorigenic potential, similarly to the use of induced pluripotent stem cells (iPSC) that are still under debate, NSC or NPC have been variably obtained from different approved tissue sources and, under specific paradigms, they have been shown as nontumorigenic and nonimmunogenic in vivo and are currently exploited in phase I and II clinical trials [12, 13].

In CNS disorders characterized by loss of neurons and glial cells, NSCs can be exploited to replace lost neurons or glial cells by transplantation of uncommitted or mature precursors through selective pre-differentiation in vitro to various stages of maturation, for example, into neuroblasts (i.e., immature neurons). Thanks to their ability to proliferate in vitro, and to their plasticity and functional stability, undifferentiated SC can be addressed toward the target phenotype in vivo by the endogenous environment and by physiological "homing" signals or can be committed in vitro to the target phenotype before transplantation.

Cell replacement might also be achieved by inducing endogenous stem cells in the adult CNS to form new neurons and glial cells. However, it has been widely demonstrated that NSC are able to modulate the tissue homeostasis in the nervous system through multiple "bystander" mechanisms other than the expected cell replacement, including the secretion of neurotrophic factors and cytokines, the clearance of toxic molecules, and modulation of either acute or chronic inflammation-driven degeneration [14, 15]. NSCs are able to release trophic factors as FGF and

EGF and cytokines [16], such as a complex array of homeostatic molecules with immune-regulatory or tissue trophic functions that ultimately reduce tissue damage and enhance endogenous repair [17].

Compelling evidence illustrated that the proliferation of endogenous NSCs is activated upon tissue damage. The lesion generates a local environment recruiting NPCs to migrate to the damaged area. Unfortunately, activated endogenous NPC are often unable to fully rescue the detrimental effects produced by the lesion [18] and the contribution by exogenous transplanted cells becomes essential.

The communication between host and stem cells also occurs through the secretion of cytokines or/and growth factors or through cellular (gap) junctional transfer of electrical, metabolic, and immunological information [19]. Previous studies also suggested that extracellular membrane vesicles might play a key role, being transferred from donor-grafted stem cells to target endogenous cells [20].

In a wide array of experimental models, there is solid evidence showing that NSCs survive after transplantation, spontaneously migrate to the lesion area, and retain their multipotency [21], where they exert therapeutic effects even when remaining undifferentiated. For example, in a rodent-induced PD or HD [22], a transplanted cell rarely gives rise to neurons, despite the improvement of the behavior. Similar results have been obtained in mice with stroke or intracerebral hemorrhage [23]. The terminal differentiation to a nonneuronal fate, as well as the propensity for maintaining undifferentiated phenotype within host tissue, support the hypothesis that transplanted NSC might be therapeutically efficacious through a bystander mechanism or an alternative to cell replacement [24]. In accordance, transplantation of GDNF-expressing hNPCs into the motor cortex of SOD1 G93A animal models has been shown effective in the rescue of both upper and lower motor neurons [25], in addition to intraspinal-injected NPCs [12, 13]. In Experimental Autoimmune Encephalomyelitis (EAE) mice, the most relevant and commonly used animal model to study autoimmune demyelinating diseases like MS [26], NSCs have shown to reduce astrogliosis and demyelination by exerting remarkable immune-regulatory and tissue-trophic effects, while still retaining mostly an undifferentiated phenotype [15].

In summary, the presence of adult NSCs in the central nervous system with the ability to proliferate and differentiate in major neural lineages and functionally integrate into neuronal circuits has prompted to their use as a potential therapeutic source. It is of fundamental importance also to determine to what extent endogenous NSCs are damaged and/or may affect the local microenvironment following neuropathogenic or traumatic events. In particular, the effort of the scientific community is currently addressed to understand if and how the NSCs respond to different pathological conditions and what are the factors that determine this response. Diseases of the central nervous system are often accompanied by an inflammatory component and an immune response, so that inflammatory molecules play a fundamental role in endogenous neurogenesis and in the subsequent differentiation of neural progenitors [27].

It has been observed that transgenic overproduction of interleukin-6 (IL-6) by astroglia is able to decrease overall neurogenesis by 63% in the hippocampal dentate gyrus of young adult mice [28]. On the other hand, other inflammatory molecules have a pleiotropic role in neurogenesis, at least for neural progenitor cells prepared from embryonic rat hippocampus [29]. Thus, the innate immune response in the brain is able to positively or negatively modulate neurogenesis in the CNS depending on the cytokines and growth factors that are predominantly expressed in the cellular environment. Recently, Peruzzotti-Jametti et al. [30] have demonstrated that transplanted NSCs ameliorate chronic CNS inflammation in MS through a reduction of succinate levels in the cerebrospinal fluid and a consequent decrease of mononuclear phagocyte infiltration, suggesting unexpected crosstalk mechanisms between NSCs and immune cells. In this scenario, it is important to note that the mechanisms involved in the regulation of innate immunity in NSC have been poorly investigated and if no treatment is currently available to arrest the progression of neuroinflammation and neurodegeneration in most of CNS disorders, this partly accounts for the poor knowledge of the "immune-like" phenotype of NSCs.

3 The Dual Role of TLR4 in CNS Homeostasis

Among TLRs, in particular TLR2 and TLR4 are present on adult NSC and NPCs, where they exert different and contrasting functions in NPC proliferation and differentiation [31, 32]. TLR2 activation, in infectious, ischemic, and inflammatory diseases could negatively impinge on brain development by inhibition of NPC proliferation [33]. On the contrary, TLR4 activation has been shown to correlate with increased proliferation of NSC/NPC after hippocampal ischemic injury [32].

TLR4 plays a key role in CNS homeostasis. Consistently, while increasing during brain development, TLR4 expression remains on adulthood, in microglia, which is a master player in neuroinflammatory processes, such as in neural cells: astrocytes, neurons, oligodendrocytes, NPC, and NSC [31, 34, 35].

With aging, a progressive and spontaneous decay of cognitive functions concurs with metabolic alterations, increase of oxidative stress, and ensuing increase of neuroinflammatory background [36]. This scenario is exacerbated in pathological situations. However, thanks to its pleiotropic effects, TLR4 activation plays a key role in neurogenesis such as in the impairment of cognitive functions and neuropathological signatures from the early development stage throughout the lifespan.

Lifelong exposure to endogenous or exogenous stressing conditions activates a cascade of responses involving both innate and adaptive immune system [37]. This sort of "preconditioning" may induce the establishment of a "physiological" state of inflammation that becomes exacerbated with aging and susceptible to the income of further stressors, thus to the activation of pathophysiological and irreversible mechanisms.

A dual role of TLR4 has been shown in the brain, where TLR4 activation may exert both pro-inflammatory and pro-neurogenic effects. Existing evidence comes predominantly from studies of in vitro and in vivo models, as well as analyses of postmortem human brain tissue and preclinical studies of TLR inhibitors. For example, after ischemic preconditioning, TLR4 mediates both the development of the inflammatory reaction and the neuroprotective "priming" of CNS [38]. In the brain, the development of neuroinflammation is related to a physiological reaction of defense against to external agents (pathogen-associated molecular patterns, PAMPs, or toxic compounds) or endogenous tissue-derived damaging molecules (danger-associated molecular patterns, DAMPS) and involves the participation of both innate and adaptive immune systems. The recruitment of the innate immune system is a very early event due to activation of microglial (immune nonneural cells resident in the CNS) and astroglial cells (neural cells activated), accompanied by the concerted activation of the adaptive immune system that involves mobilization of immune cells from the bone marrow and infiltration of monocyte cells like T-lymphocytes and macrophages through a transient permeabilization of the blood–brain barrier (BBB). Under pathological circumstances, the neuroinflammatory reaction overcomes the physiological threshold and becomes detrimental to axonal function and neurogenesis.

Accordingly, inflammaging, that is, the chronic inflammatory process characterized by an imbalance of pro- and anti-inflammatory mechanisms, has been recognized as operative in several age-related and notably neurodegenerative diseases. Inflammaging is part of the complex adaptive mechanisms ("remodeling") that are ongoing through the lifespan, and which function to prevent or mitigate endogenous processes of tissue disruption and degenerative change(s). The absence of an adequate anti-inflammatory response can foster inflammaging, which propagates on both local (i.e., from cell to cell) and systemic levels (e.g., via exosomes and other molecules present in the blood). In general, this scenario is compatible with the hypothesis that inflammaging represents a hormetic or hormetic-like effect, in which low levels of inflammatory stress may prompt induction of anti-inflammatory mediators and mechanisms, while sustained pro-inflammatory stress incurs higher and more durable levels of inflammatory substances, which, in turn, prompt a local-to-systemic effect and more diverse inflammatory response(s) [39]. In the development of a neuroinflammatory background, TLR4 is differentially involved dependently on the early or advanced stages of the process.

A persistent acute neuroinflammation can turn to a chronic neuroinflammation as it accumulates damage, bringing about neuronal degeneration. The outcome of neuroinflammation correlates with the time span of the inflammatory response and the activation state of microglial [40] and astroglial cells which play concerted and complementary roles in the defense of the brain.

Microglial cells represent the primary innate immunological cell type in the brain. While changing from a stellate surveillant to an amoeboid reactive phenotype and producing pro-inflammatory signals upon insult and/or pathological input [41], microglial cells are also able to promote NPC proliferation, scavenging of cell debris, and secretion of neurotrophic factors. A new concept is that neural cells actively participate to the neuroinflammatory process with their intrinsic innate immunity, by releasing pro-inflammatory cytokines, chemokines, and reactive oxygen species. Hence, while able to promote neurogenesis and neuronal development under physiological conditions, astrocytes may acquire a fibrotic reactive phenotype upon activation by intrinsic or extrinsic damage signals and become neurotoxic and pathogenic [42]. Both the loss of the beneficial effects and hyperactivation of astrocytes have been suggested as contributing to an increased neuronal susceptibility and altered metabolism [43], with increased levels of cytosolic metabolites that can cause oxidative stress and disruption of mitochondrial pattern and efficiency. The release of pro-inflammatory mediators in response to neural dysfunction may be helpful, neutral or even deleterious to normal cellular survival, with TLR4 playing a multifaceted role in the development of specific environmental conditions or pathological contexts.

TLR4 is the primary signaling receptor for gram-negative bacteria lipopolysaccharide (LPS). The activation of TLR4 pathway by LPS requires the participation of the LPS-binding proteins LBP and soluble and membrane-bound CD14 receptor and the adaptor MD-2, which associates noncovalently with TLR4 to form the activated homodimer $(LPS/MD\text{-}2/TLR4)_2$ on the cell membrane that starts the intracellular signal leading to NF-kB activation and pro-inflammatory cytokine production [44]. The homodimer transmits the signal downstream through two distinct and consecutive pathways. One pathway starts from the cell membrane complex by the recruitment of myeloid differentiation primary response gene 88 (MyD88) and adapter myelin and lymphocyte protein (MAL) (MyD88-dependent pathway). A supramolecular signaling complex formed by six molecules of MyD88 is formed (Myddosome) that induces NF-κB activation and the production of a number of pro-inflammatory proteins [45]. Once engaged by mCD14, TLR4-MD2 undergoes an internalization process and moves in the endosomal compartment, where it triggers the TRIF Related Adaptor Molecule (TRAM) and TIR-Domain-Containing Adapter-Inducing Interferon-β (TRIF)-dependent pathway, that sustains the activation of NF-κB and also induces the production of type I interferons (IFNs) [46].

The activation of TLR4 by endogenous DAMPs that very likely occurs in CNS, has been less characterized from a structural point of view, and the roles of the co-receptor CD14 and adaptor MD-2 have to be clarified. On the other hand, there is increasing pharmacological interest in TLRs targeting in CNS pathologies [47].

Chemical agents can inhibit PAMP- and DAMP-TLR4 signaling through a direct interaction outside the cell with the extracellular domain of TLR4, or inhibit the interaction of TLR4 with MD-2 and CD14. TLR4 antagonists can alternatively enter into the cell and impair MyD88 or TRAM/TRIF signal pathways. In this view, the design of different TLR4 antagonists targeting different molecules and with different capacity to penetrate the cell membrane (and other tissue barriers) may be crucial to tailor specific therapeutic strategies to approach neuroinflammation and neuroregeneration in CNS disorders.

4 The Role of TLR4 in Neurogenesis Through Regulating Neural Stem Cells (NSC) and Neural Progenitor Cells (NPC)

While notably protecting from inflammation in several systems [48, 49], recent evidences have shown that TLR4 deficiency impairs oligodendrocyte formation after spinal cord injury [50]. In a similar fashion, while MSRV multiple sclerosis–associated retrovirus (MSRV) envelope protein, a potent agonist of TLR4 has been shown to impair differentiation of oligodendrocyte precursors to mature oligodendrocytes [51], LPS activation of TLR4 has been shown to promote remyelination [52], indicating that TLR4 signaling has to be finely tuned in order to promote neuroregeneration rather than neurodegeneration. Interestingly, in humans, TLR4 (D299G) mutation causing impairment of LPS signaling does not affect human or experimental sepsis caused by polymicrobial infection, suggesting that also other immune receptors may compensate for TLR4 defects [53].

TLR4 is known to be a master player in the development of neuroinflammation processes and to be upregulated in ALS animal models. Indeed, inhibition of TLR4 by small molecule antagonists has been shown to variably attenuate neuroinflammation development in ALS animal models [54, 55]. In a similar fashion, intraspinal transplantation of hNSCs in ALS rats has been shown to ameliorate disease progression through immunomodulatory effects. Previous studies in murine cells or animal models have shown thatTLR4 plays multiple and controversial roles in neurogenesis [56], but the lack of a human system to study the CNS and the paucity of data on human patients have represented a roadblock to the appropriate knowledge of some pathophysiological mechanisms and to plan possible therapeutic strategies.

In a translational perspective, the investigation of the role of TLR4 in the regulation of hNSC lines promoted as a tool for cell-mediated therapy of aging disorders is of utmost importance, suggesting TLR4 as a novel target either for the identification of novel biomarkers for rare diseases or future therapeutic approaches. In immune-mediated experimental demyelination – both in rodents and nonhuman primates – transplanted NPC have been shown to possess an inherent and inducible ability to mediate efficient "bystander" myelin repair and axonal rescue. In particular, it seems that undifferentiated NPCs display the higher "bystander" therapeutic potential and, once transplanted, are regulated by both CNS-resident and blood-borne inflammatory cells releasing in situ major stem cell regulators. The immunomodulatory capacity of exogenous NPCs together with the formation of the atypical ectopic perivascular niches provide an example of reciprocal crosstalk between the inflamed microenvironment(s) and transplanted therapeutic NPCs. Given this perspective, new treatments of neurodegenerative disease sharing the common hallmark of inflammaging may be envisioned that strategically are aimed at exerting hormetic effects to sustain anti-inflammatory responses, inclusive perhaps, of facilitating the immunomodulatory effect of the stem cell-mediated therapeutic approach.

We have then investigated [35] the role of TLR4 in regulating hNSC properties by analyzing in vitro the effects of long-term TLR4 stimulation by LPS and inhibition by the synthetic glycolipid FP7 active as antagonist [57, 58]. TLR4 antagonism by FP7 is based on strong binding to MD-2 and CD14, and subsequent inhibition of formation of the final activated dimer TLR4/MD-2 [57, 58] has an impact on the proliferation dynamics and cell fate of human NSCs. We have shown that hNSC do express TLR4, its adaptor MD-2, and the co-receptor CD14 on the cell membrane. Importantly, our findings indicate that treatment with LPS exerts slight increase of hNSC proliferation, while more relevant effects are evident by using the synthetic antagonist, suggesting that TLR4 signaling is active in hNSC under basal conditions in vitro and TLR4 signal is required for cell proliferation and survival. TLR4 activation by LPS has a positive effect on hNSC proliferation and long-term survival, while TLR4 inhibition by the synthetic antagonist remarkably decreases the proliferation rate and cell viability in a time- and dose-dependent fashion.

TLR4 stimulation by LPS during differentiation enhances neuronogenic potential of hNSC and favors both neuronal and oligodendroglial survival, while TLR4 inhibition by synthetic antagonist FP7 accelerates the differentiation process by fostering hNSC to the transient amplifying neural progenitor (NPC) stage, finally leading to yield a lower percentage of neuronal and oligodendroglial cells [35].

Interestingly, the alteration of mitochondrial and lysosomal patterns in hNSC by chemical inhibition of TLR4 signaling correlate with the apoptotic marker Caspase3, suggesting that excessive downregulation of TLR4 in hNSC leads to oxidative stress and metabolic dysfunction in a fashion comparable to a pro-inflammatory situation. In a similar dual interpretation, we observed some activation of the inflammasome pathway indicating that TLR4 may participate in preconditioning of hNSC vulnerability and senescence [35].

We have provided evidence of TLR4 expression in hNSC in vivo 40 days after transplantation into the spinal cord of SOD1 G93A rats. These results confirm that TLR4 may play hormetic-like effect also in neurogenesis: while the TLR4 overexpression in the spinal cord of SOD1 rats during ALS progression is likely involved in the development of the neuroinflammation, TLR4 basal expression and activation is needed in hNSC to favor their engraftment. We also demonstrated that clinical-grade hNSC retain TLR4 expression after transplantation into the brain of nude SCID mice, thus irrespectively both of the site of injection and of the neuroinflammatory background. To note, hNSC retaining TLR4 expression after transplantation co-express nestin consistently with a role of TLR4 in fostering survival and proliferation of a stem-like cell [35]. In the actual perspective of stem cell–mediated therapy of neurodegenerative diseases like ALS, the analysis of TLR4-related molecules/proteins along with disease development and progression appears of essence for the identification of novel diagnostic and prognostic biomarkers. Nonetheless, novel therapeutic strategies based on TLR4 modulation should be tailored according to these observations.

Consistently with our hypothesis on TLR4 effects on exogenous hNSCs after transplantation, Palma-Tortosa et al. [59] have shown that endogenous NSC are actively stimulated to proliferate by TLR4 activation after stroke, thus confirming the relevant role of TLR4 in promoting neurogenesis under non-physiological conditions. Interesting, a recent study by Muneoka S. et al. [60] has demonstrated that

TLR4 on circumventricular NSCs in the adult mouse brain functions as a central regulator for thermogenesis under inflamed and normal conditions through mediating response to LPS-induced inflammatory stimulus.

Altogether, these results indicate that TLR4 activity has not to be turned on or off, but needs being finely tuned in order to promote neuroregeneration rather than neurodegeneration and to mediate the development of healing immunomodulation rather than of detrimental neuroinflammation.

References

1. Weiss S, Reynolds BA, Vescovi AL, Morshead C, Craig CG, et al. Is there a neuronal steam cell in the mammalian forebrain? Trends Neurosci. 1996a;19:387393.
2. Reynolds BA, Weiss S. Generation of neurons and astrocytes from isolated cells of the adult mammalian central nervous system. Science. 1992;255:1707–10.
3. Roy NS, Wang S, Jiang L, Kang J, Benraiss A, et al. In vitro neurogenesis by progenitor cells isolated from the adult human hippocampus. Nat Med. 2000;6:271–7.
4. Weiss S, Dunne C, Hewson J, Wohl C, Wheatley M, et al. Multipotent CNS steam cells are present in the adult mammalian spinal cord and ventricular neurotaxis. J Neurosci. 1996b;16:7599–609.
5. Lazutkin A, Podgorny O, Enikolopov G. Modes of division and differentiation of neural stem cells. Behav Brain Res. 2019;18(374):112118.
6. Ma DK, Ming GL, Song H. Glial influences on neural stem cell development: cellular niches for adult neurogenesis. Curr Opin Neurobiol. 2005;15:514–20.
7. Alvarez-Buylla A, Lim DA. For the long run: maintaining germinal niches in the adult brain. Neuron. 2004;41:683–6.
8. Song H, Stevens CF, Gage FH. Astroglia induce neurogenesis from adult neural stem cells. Nature. 2002;417:39–44.
9. Ma DK. Kim WR. Song H. Activity-dependent extrinsic regulation of adult hippocampal and olfactory bulb neurogenesis. Ann N Y Acad Sci: Guo LM; 2009. in press
10. Pluchino S, Cusimano M, Bacigaluppi M, Martino G. Remodelling the injured CNS through the establishment of atypical ectopic perivascular neural stem cell niches. Arch Ital Biol. 2010;148:173–83.
11. Donega V, Van Velthoven CTJ, Nijboer CH, Van Bel F, Kas MJH, et al. Intranasal mesenchymal stem cell treatment for neonatal brain damage: long-term cognitive and sensorimotor improvement. PLoS One. 2013;8(1):e51253.
12. Glass JD, Hertzberg VS, Boulis NM, Riley J, Federici T. Transplantation of spinal cord-derived neural stem cells for ALS: analysis of phase 1 and 2 trials. Neurology. 2016;87(4):392–400.
13. Mazzini L, Gelati M, Profico DC, Sgaravizzi G, Projetti Pensi M, et al. Human neural stem cell transplantation in ALS: initial results from a phase I trial. J Transl Med. 2015;13:17.
14. Pluchino S, Quattrini A, Brambilla E, Gritti A, Salani G, et al. Injection of adult neurospheres induces recovery in a chronic model of multiple sclerosis. Nature. 2003;422:688–94.
15. Pluchino S, Gritti A, Blezer E, Amadio S, Brambilla E, et al. Human neural stem cells ameliorate autoimmune encephalomyelitis in non-human primates. Ann Neurol. 2009;66(3):343–54.
16. Bottai, et al. Neural stem cells in the adult nervous system. J Hematother Stem Cell Res. 2003;12(6):655–70.
17. Li L, Xie T. Stem cell niche: structure and function. Annu Rev Cell Dev Biol. 2005;21:605–31.
18. Arvidsson Collin T, Kirik D, Kokaia Z, Lindvall O. Neuronal replacement from endogenous precursors in the adult brain after stroke. Nat Med. 2002;8:963–70.

19. Pluchino S, Cossetti C. How stem cells speak with host immune cells in inflammatory brain diseases. Glia. 2013;61(9):1379–401.
20. Cossetti C, Iraci N, Mercer TR, Leonardi T, Alpi E, et al. Extracellular vesicles from neural stem cells transfer IFN-γ via Ifngr1 to activate Stat1 signaling in target cells. Mol Cell. 2014;56(2):193–204.
21. Goutman SA, Chen KS, Feldman EL. Recent advances and the future of stem cell therapies in amyotrophic lateral sclerosis. Neurotherapeutics. 2015;12(2):428–48.
22. Kordower JH. Introduction to Parkinson's and Huntington's disease. CNS Regen. 1999;10:295–8.
23. Lee KR, Kawai N, Kim S, Or Sagher BA, Julian T. Mechanisms of edema formation after intracerebral hemorrhage: effects of thrombin on cerebral blood flow, blood-brain barrier permeability, and cell survival in a rat model. J Neurosurg. 1997;86:272–8.
24. Martino G, Pluchino S. Neural stem cells: guardians of the brain. Nat Cell Biol. 2007;9(9):1031–4.
25. Thomsen GM, Avalos P, Ma A, Alkaslasi M, Cho N, et al. Transplantation of Neural Progenitor Cells Expressing Glial Cell Line-Derived Neurotrophic Factor into the Motor Cortex as a Strategy to Treat Amyotrophic Lateral Sclerosis. Stem Cells. 2018;36:1122–31.
26. Howland D, Liu J, She Y, Goad B, Maragakis N, et al. Focal loss of the glutamate transporter EAAT2 in a transgenic rat model of SOD1 mutant-mediated amyotrophic lateral sclerosis (ALS). Proc Natl Acad Sci U S A. 2002;99:1604–9.
27. Simard AR, Rivest S. Role of inflammation in the neurobiology of stem cells. Neuroreport. 2004;15:2305–10.
28. Vallieres L, Campbell IL, Gage FH, Sawchenko PE. Reduced hippocampal neurogenesis in adult transgenic mice with chronic astrocytic production of interleukin-6. J Neurosci. 2002;22:486–92.
29. Harada J, Foley M, Moskowitz MA, Waeber C. Sphingosine-1- phosphate induces proliferation and morphological changes of neural progenitor cells. J Neurochem. 2004;88:1026–39.
30. Peruzzotti-Jametti L, Bernstock JD, Vicario N, Costa ASH, Kwok CK, et al. Macrophage-derived extracellular succinate licenses neural stem cells to suppress chronic Neuroinflammation. Cell Stem Cell. 2018;22(3):355–68.
31. Rolls A, Shechter R, London A, Ziv Y, Ronen A, Levy R, et al. Toll-like receptors modulate adult hippocampal neurogenesis. Nat Cell Biol. 2007;9:1081–8.
32. Ye YZX, Li Z, Jia Y, He XHJ. Association between toll-like receptor 4 expression and neural stem cell proliferation in the hippocampus following traumatic brain injury in mice. Int J Mol Sci. 2014;15:12651–64.
33. Okun E, Griffioen KJ, Lathia JD, Tang SC, Mattson MP, et al. Toll-like receptors in neurodegeneration. Brain Res Rev. 2009;59:278–92.
34. Covacu R, Arvidsson L, Andersson A, Khademi M, Erlandsson-Harris H, et al. TLR activation induces TNF-α production from adult neural stem/progenitor cells. J Immunol. 2009;182(11):6889–95.
35. Grasselli C, Ferrari D, Zalfa C, Soncini M, Mazzoccoli G, et al. Toll-like receptor 4 modulation influences human neural stem cell proliferation and differentiation. Cell Death Dis. 2018;9:280.
36. Logan S, Royce GH, Owen D, Farley J, Ranjo-Bishop M, et al. Accelerated decline in cognition in a mouse model of increased oxidative stress. GeroScience. 2019;41(5):591–607.
37. Ottaviani E, Franceschi C. A new theory on the common evolutionary origin of natural immunity, inflammation and stress response: the invertebrate phagocytic immunocyte as an eyewitness. Domest Anim Endocrinol. 1998;15(5):291–6.
38. Pradillo JM, Fernández-López D, García-Yébenes I, Sobrado M, Hurtado O, et al. Toll-like receptor 4 is involved in neuroprotection afforded by ischemic preconditioning. J Neurochem. 2009;109(1):287–94.
39. Calabrese V, Santoro A, Monti D, Crupi R, Di Paola R, et al. Aging and Parkinson's disease: Inflammaging, neuroinflammation and biological remodeling as key factors in pathogenesis. Free Radic Biol Med. 2018;115:80–91.

40. Yang H, Yang S, Huang S, Tang B, Guo J. Microglial activation in the pathogenesis of Huntington's disease. Front Aging Neurosci. 2017;9:193.
41. Lull ME, Block ML. Microglial activation and chronic neurodegeneration. Neurotherapeutics. 2010;7(4):354–65.
42. Zalfa C, Verpelli C, D'Avanzo F, Tomanin R, Vicidomini C, Cajola L, et al. Oxidative damage with glial degeneration drives neuronal demise in MPSII disease. Cell Death Dis. 2016;7:2331.
43. Fusar-Poli P, Meyer-Lindenberg A. Striatal presynaptic dopamine in schizophrenia,PartII: meta-analysis ofF-18/C-11-DOPA PETstudies. Schizophr Bull. 2013;39:33–42.
44. Park BSSD, Kim HM, Choi BSLH, Lee JO. The structural basis of lipopolysaccharide recognition by the TLR4-MD-2 complex. Nature. 2009;458:1191–5.
45. Lin SC, Lo YC, Wu H. Helical assembly in the MyD88-IRAK4-IRAK2 complex in TLR/IL-1R signalling. Nature. 2010;465(7300):885–90.
46. Fitzgerald KA, Rowe DC, Barnes BJ, Caffrey DR, Visintin A, et al. LPS-TLR4 signaling to IRF-3/7 and NF-kappaB involves the toll adapters TRAM and TRIF. J Exp Med. 2003;198(7):1043–55.
47. Anstadt EJ, Fujiwara M, Wasko N, Nichols F, Clark RB. TLR tolerance as a treatment for central nervous system autoimmunity. J Immunol. 2016;197(6):2110–8.
48. Ghosh AK, O'Brien M, Mau T, Yung R. Toll-like receptor 4 (TLR4) deficient mice are protected from adipose tissue inflammation in aging. Aging (Albany NY). 2017;9(9):1971–82.
49. Weber SN, Bohner A, Dapito DH, Schwabe RF, Lammert F. TLR4 deficiency protects against hepatic fibrosis and Diethylnitrosamine-induced pre-carcinogenic liver injury in fibrotic liver. PLoS One. 2016;11(7):e0158819.
50. Church J, Kigerl K, Lerch J, Popovich P, McTigue D. TLR4 deficiency impairs oligodendrocyte formation in the injured spinal cord. J Neurosci. 2016;36(23):6352–64.
51. Madeira A, Burgelin I, Perron H, Curtin F, Lang AB, Faucard R. MSRV envelope protein is a potent, endogenous and pathogenic agonist of human toll-like receptor 4: relevance of GNbAC1 in multiple sclerosis treatment. J Neuroimmunol. 2016;291:29–38.
52. Glezer I, Lapointe A, Rivest S. Innate immunity triggers oligodendrocyte progenitor reactivity and confines damages to brain injuries. FASEB J. 2006;20:750–2.
53. Feterowski C, Emmanuilidis K, Miethke T, Gerauer K, Rump M, Ulm K, et al. Effects of functional toll-like receptor-4 mutations on the immune response to human and experimental sepsis. Immunology. 2003;109(3):426–31.
54. Fellner A, Barhum Y, Angel A, Perets N, Steiner I, et al. Toll-like Receptor-4 inhibitor TAK-242 attenuates motor dysfunction and spinal cord pathology in an amyotrophic lateral sclerosis mouse model. Int J Mol Sci. 2017;18(8):1666.
55. De Paola M, Sestito SE, Mariani A, Memo C, Fanelli R, et al. Synthetic and natural small molecule TLR4 antagonists inhibit motoneuron death in cultures from ALS mouse model. Pharmacol Res. 2016;103:180–7.
56. Zeuner M, Bieback K, Widera D. Controversial role of toll-like receptor 4 in adult stem cells. Stem Cell Rev. 2015;11:621–34.
57. Cighetti R, Ciaramelli C, Sestito SE, Zanoni I, Kubik Ł, et al. Modulation of CD14 and TLR4. MD-2 activities by a synthetic lipid a mimetic. Chembiochem. 2014;15:250–8.
58. Facchini FA, Zaffaroni L, Minotti A, Rapisarda S, Calabrese V, et al. Structure-activity relationship in monosaccharide-based toll-like receptor 4 (TLR4) antagonists. J Med Chem. 2018;61(7):2895–909.
59. Palma-Tortosa S, Hurtado O, Pratillo JM, Ferreras-Martìn R. Garcià-Yébenes et al. toll-like receptor 4 regulates subventricular zone proliferation and neuroblast migration after experimental stroke. Brain Behav Immun. 2019;80:573–82.
60. Muneoka S, Murayama S, Nakano Y, Miyata S. TLR4 in circumventricular neural stem cells is a negative regulator for thermogenic pathways in the mouse brain. J Neuroimmunol. 2019;331:58–73.

Toll-Like Receptor 4 and the World of microRNAs

Monica Molteni and Carlo Rossetti

Abstract Recent studies have highlighted the importance of microRNAs (miRNAs) in the fine-tuning of cellular response after TLR4 activation, in the timely coordination of both inflammatory and resolution phases of the response. In this chapter, we examine the intracellular role of TLR4–miRNA axis in immune and nonimmune cells and the importance of miRNAs released in exosomes, after TLR4 triggering, for the crosstalk among different cellular subsets.

Keywords Toll-Like Receptor 4 (TLR4) · microRNAs · Inflammation · Immune cells · Nonimmune cells · Exosomes

1 Introduction

Toll-Like Receptor 4 (TLR4) is a key receptor involved in the inflammatory response induced by pathogens (mainly gram-negative bacteria) or by tissue injury [1]. Interaction with specific agonists triggers a signaling cascade that activates downstream transcription factors, such as MyD88-dependent NF-κB, AP-1, and MyD88-independent IRF3, that are ultimately responsible for the initiation of the proinflammatory response needed to eradicate infection and to restore tissue integrity [2]. As in several other biological processes, TLR4 activation involves feedback mechanisms with the aim of controlling the intensity of proinflammatory activation, finely directing not only the timing of the response but also the switch toward the resolution phase involving tissue repairing processes. Among these mechanisms, the timely coordinated production of specific microRNAs (miRNAs) by TLR4 triggering have received considerable attention as newly identified regulators playing central roles both in the control of intracellular responses and in the communications among cells that belong or not to the immune compartment. Increasing evidence indicates that miRNAs have a critical function as regulators of hematopoiesis and immune cell differentiation but also significantly influence the outcome of the

M. Molteni (✉) · C. Rossetti
Department of Medicine and Surgery, University of Insubria, Varese, Italy
e-mail: monica.molteni@uninsubria.it

C. Rossetti, F. Peri (eds.), *The Role of Toll-Like Receptor 4 in Infectious and Non Infectious Inflammation*, Progress in Inflammation Research 87,
https://doi.org/10.1007/978-3-030-56319-6_10

immune responses to infection [3]. Moreover, being TLR4 expressed also on the plasma membrane of nonimmune cells, including epithelial, endothelial, tumor, and mesenchymal stem cells, it is evident that TLR4-miRNA regulation could have a more relevant role than previously thought, as suggested also by the increasing number of papers recently published on this topic.

2 MiRNAs: Canonical Biogenesis Pathway and Mechanisms of Action

MiRNAs are a class of noncoding single-stranded RNA molecules (20–24 nucleotides in length) having a cytoplasmic post-transcriptional regulatory role on protein-coding gene expression [4, 5]. In humans, over 2000 miRNAs have been identified as true mature miRNAs [6, 7], and it has been hypothesized that they could influence the output up to 60% of protein-coding genes [4, 5, 8].

Transcription of miRNA genes is essentially similar to that observed for protein-coding genes, undergoing the same activation and regulatory mechanisms [6, 9]. MiRNAs are transcribed from DNA sequences into quite long primary miRNAs (pri-miRNA) by RNA polymerase II and then cleaved in the nucleus into pre-miRNAs (60–70 nucleotide hairpin intermediates) by the microprocessor complex, consisting of ribonuclease III enzyme Drosha and RNA binding protein DiGeorge Syndrome Critical Region 8 (DGCR8) [10–12]. Once a pre-miRNA is produced, it is actively exported to the cytoplasm, where it is processed by the RNase III endonuclease Dicer, resulting in a mature miRNA duplex [10, 13, 14] that is loaded into the Argonaute (AGO) proteins (AGO1 to AGO4 in humans). Only one strand of the duplex (either 5′ or 3′ strands based on the thermodynamic stability of the two ends of the duplex) remains as mature miRNA to form the RNA-induced silencing complex (RISC); the other strand is degraded. The RISC represents the effector complex recognizing short sequences in the 3′ untranslated region (UTR) of the target messenger RNAs (mRNAs) [14].

The main mechanisms of miRNA-mediated post-transcriptional regulation depends on the grade of sequence complementarity between miRNA and target mRNA and at least involve:

1- Direct AGO2-catalyzed mRNA cleavage if miRNA "seed region" (e.g., nucleotides 2–8) base pairing with mRNA in the 3′-UTR is perfect [5, 15].
2- Translation repression pathways by blocking the initiation step [5, 16] and mRNA destabilization by de-adenylation and de-capping with final mRNA degradation, eventually occurring in P-bodies [17, 18].

To add further complexity to the miRNA regulatory effects, it has been recently demonstrated that miRNAs can target promoter elements in the nucleus eventually also showing gene-activating functions [8, 19].

3 TLR4–miRNA Axis in the Innate Immune Response

The first documented relationship between TLR4 triggering by lipopolysaccharide (LPS) and miRNA induction was found by Baltimore's research group in 2006 [20]. In this paper, Taganov and colleagues for the first time demonstrated that miR-146a, miR-146b, miR-132, and miR-155 are LPS-responsive genes in human monocytic THP-1 cells. After this, several other papers have been published showing the role of many other miRNAs in the context of TLR4 activation and cell signaling in innate immune cells. In particular, it is possible to distinguish miRNAs directly induced or suppressed by TLR4 triggering, from those targeting TLR4 expression and/or signaling proteins (Table 1). In this chapter, only the miRNAs whose function has been clearly demonstrated by independent research groups are presented.

3.1 *MiRNA Upregulated by TLR4 Triggering*

3.1.1 The miR-146 Family: miR-146a-5p and miR-146b-5p

MiR-146a-5p and miR-146b-5p (here called miR-146a and miR-146b) are two evolutionary conserved miRNAs, located on separate chromosomes and differing in their sequences for only two nucleotides at the 3′ end, suggesting that they could share the same target mRNAs. It has been demonstrated that human monocytic

Table 1 MiRNAs regulated by TLR4 triggering and miRNAs targeting TLR4 expression and/or protein signaling

	MiRNAs regulated by TLR4 triggering	
miRNA	**Verified targets**	**Reference**
Upregulated		
miR-146a	IRAK1, IRAK2, TRAF6, TLR4	[20, 23, 25, 105]
miR-146b	MyD88, IRAK1, TRAF6, TLR4	[31, 32]
miR-155	IKKβ, IKKε, FADD, TAB2, p38 MAPK, TNF, SHIP1, SOCS1, BCL6	[37, 41–46]
miR-21	MyD88, IRAK1, PDCD4, PTEN, IL-12	[63, 64]
miR-132	IRAK4, acetylcholinesterase	[67–69]
Downregulated		
miR-223	IKKα, STAT3, NLRP3, TLR4	[71, 73, 76–78]
miR-125b	TNFα, MIP-1α, BIK, MTP18	[80, 81, 83]
	miRNAs targeting TLR4 expression and/or protein signaling	
miRNA	**Verified targets**	**Reference**
Let-7i, e, a, b	TLR4	[79, 84–86]
miR-511	TLR4	[87, 88]
miR-200b, c	MyD88, c-Jun MAPK	[89, 90]

THP-1 cells incubated with bacterial LPS rapidly increase transcription of both miR-146a and miR-146b, but only mature miR-146a is observed in the cytoplasm [20, 21]. This indicates that, even if they share the same "seed region," miR-146a and miR-146b could also have different nonredundant biological functions [3]. Further studies confirmed upregulation of miR-146a in primary monocyte/macrophages, dendritic cells, and microglial cells exposed to LPS [22–24]. The kinetics of miR-146a expression, after LPS incubation, is characterized by a gradual increase over 24 h both in THP-1 cells and in monocytes [21, 23]. Analysis of miR-146a promoter showed transcriptional NF-κB binding sites; thus LPS-induced NF-κB activation directly results in miR-146a upregulation [20]. It has been proposed that miR-146a functions as a negative regulator of inflammation [3]. Several experimental results support this statement. Interleukin-1 receptor-associated kinases (IRAK1, IRAK2) and TNF receptor-associated factor 6 (TRAF6) are key adapter proteins recruited after TLR4 triggering which act upstream NF-κB transcription factor [1]. It has been shown that IRAKs and TRAF6 gene expressions are post-translationally repressed by miR-146a [20, 25]. Furthermore, miR146a is critically involved in the process of endotoxin tolerance, a phenomenon in which monocytes/macrophages display reduced capacity to respond to repeated stimulation with LPS, showing reduced production of proinflammatory mediators, such as tumor necrosis alpha (TNF-α) [26]. Endotoxin tolerance has been suggested to be a means to avoid hyperinflammation due to repeated stimulation with LPS. Nahid and colleagues [21] identified in the miR-146a one of the key regulators of this process. In particular, they showed miR-146a overexpression in vitro during the tolerized state of THP-1cells. Moreover, transfection of cells with miR-146a was shown to mimic LPS priming, whereas transfection with antagomir abolished endotoxin tolerance. Concomitantly, a decrease of IRAK1 and TRAF6 protein expression was observed, correlating with the dose of LPS used for priming. An important role for miR-146a has also been observed in the mechanism of cross-tolerance induced by a nonconventional LPS obtained from a cyanobacterium having TLR4/MD2 antagonist activity [23]. In vivo contribution of miR-146a to endotoxin tolerance has been demonstrated in the physiological tolerance to intestinal bacteria observed in neonates [27] and in sepsis [28]. MiR-146a has been found within exosomes released by dendritic cells and, following uptake in recipient cells, it mediates target repression both in vitro and in vivo [29]. In mutant mice with targeted deletion of miR-146a, macrophages were hyperresponsive to LPS. Aging miR-146a-null mice were characterized by loss of peripheral T lymphocyte tolerance and developed an autoimmune disorder with splenomegaly, lymphadenopathy, multiorgan inflammation, and premature death [30].

Recently, beyond miR146a, also mature miR146b has been implicated in the response of human monocytes to LPS stimulation and in endotoxin tolerance in vitro [31, 32]. The increased production of miR146b was not direct as for miR146a but was dependent on IL-10 autocrine signaling pathway activated after LPS challenge in primary monocytes. The partial discrepancy with the previous observations of Taganov and colleagues [20] probably relies on the fact that THP1 monocytic cells, employed in Taganov experiments, produce very low amount of

IL-10 in culture supernatants in comparison with human primary monocytes [23]. Interestingly, overexpression of miR-146b by cell transfection showed a reduction of Myeloid Differentiation 88 (MyD88), IRAK1, and TRAF6 protein expressions. Furthermore, TLR4 itself was shown to be a target of miRNA-146b, by blocking translation of TLR4 mRNA [32].

3.1.2 MiR-155-5p

MiR-155 is highly expressed in thymus and spleen and has been demonstrated to have an important role in the development and function of both innate and adaptive immune cells [33–35]. MiR-155 is rapidly induced in response to infection or tissue injury, driving the proinflammatory response. Exposure of monocyte/macrophages, THP1 cells and microglia to LPS results in the activation of miR-155 production in vitro [20, 24, 36, 37]. Interestingly the kinetics of miR-155 production, after LPS challenge, is more rapid than that observed for miR146a [23]. It has been demonstrated that miR-146a and miR-155 are coordinately regulated during macrophage proinflammatory activation and endotoxin tolerance, representing, respectively, negative and positive regulatory elements of NF-κB transcription factor, both in vitro and in vivo [38, 39]. In miR-146a-deficient mice, elevated miR-155 expression potentiates NF-κB activity, inducing increased proinflammatory response, thus indicating a dominant role of miR-155 in promoting inflammation [38]. Transcription of miR-155 gene is under control of several transcription factors, including NF-κB, Interferon Regulatory Factors (IRF), Interferon-Sensitive Response Elements (ISRE), and AP-1 [40]. MiR-155 mRNA targets appear to be mainly products acting as inhibitors of the innate immune response. Among experimentally verified targets of miR-155 action, there are phosphatidylinositol-3,4,5-triphosphate 5-phosphatase 1 (SHIP1), which inhibits the TLR/PI_3/AKT kinase pathway, suppressor of cytokine signaling 1 (SOCS1), and B-cell lymphoma 6 protein (BCL6), which negatively regulates NF-κB transcription factor [37, 41–44]. Furthermore, miR-155 induced by LPS treatment has been shown to repress mRNAs of some proteins involved in TLR4 signaling pathway, such as those coding for IKKε and IKKβ kinases and TAB2 [45, 46]. In mouse macrophages and Kupffer cells, miR-155 upregulation following LPS treatment has been documented to increase TNF-α secretion by favoring mRNA stability, even though the mechanism is unknown [45, 47].

It has been hypothesized that the induction of miR-155 could represent a way for the immune system to rapidly expand the myeloid cell population during inflammation [48] and effectively miR-155 overexpression in hematopoietic stem cells in mice have been shown to cause myeloproliferative disorders [48]. MiR-155 also represents a critical element at the interface of innate and adaptive immunity. M1 macrophage polarization process is associated with increased expression of miR-155 [49], as well as dendritic cells overexpression of miR-155 alone is enough to break tolerance in mice [50]. In addition, miR-155-deficient mice cannot be successfully immunized against *Salmonella typhimurium*; analysis of dendritic cell function showed their inability to trigger T-cell activation after antigen presentation [51].

Pathogen-specific and tumor-specific CD8+ T-cell response has been demonstrated to be critically linked to miR-155, which has been shown to specifically restrain immune senescence and functional T-cell exhaustion, thus potentiating adaptive immune response [52, 53].

In human diseases characterized by pathological proinflammatory processes, such as rheumatoid arthritis (RA), the expression of miR-155 in monocytes was found higher in RA patients than in healthy controls; in particular, miR-155 was upregulated in synovial tissue macrophages, and, in experimental models of RA, miR-155-deficient mice were resistant to collagen-induced arthritis [54, 55].

3.1.3 MiR-21

TLR4 triggering by LPS in monocyte/macrophages stimulates the production of mature miR-21 in an NF-κB-dependent manner, mediated by the myD88 pathway [56]. It has been shown that miR-21 has an anti-inflammatory action by acting as repressor of the proinflammatory tumor suppressor programmed cell death 4 (PDCD4) and the phosphatase and tensin homologue PTEN [56–58]. The inhibition of PDCD4 expression by miR-21 is responsible for increasing IL-10 production, thus promoting a negative feedback loop controlling inflammation. MiR-21 upregulation has been implicated in the resolution of inflammation after an injury by promoting the efferocytosis of apoptotic cells by monocyte-derived macrophages. In detail, LPS-treated macrophages incubated with apoptotic cells showed enhanced miR-21 expression [59].

The role of miR-21 was investigated in vivo in models of peritonitis induced by LPS [60]. In these experiments, peritoneal macrophage expression of miR-21 was increased after LPS treatment but was delayed until 48 hours after cecal ligation and puncture. Furthermore, the survival of miR21-null mice was decreased after LPS-induced peritonitis [60] MiR-21 dysregulation has been involved in pathological conditions (reviewed in 57); several studies have demonstrated a key role of miR-21 in the interplay between innate immune cells (tumor-associated macrophages) and tumor environment. MiR-21 deficiency has been shown to confer enhanced antitumor immunity by favoring M1 polarization of macrophages [61, 62]. MiR-21-deficient DCs treated with LPS showed increase production of proinflammatory IL-12 [63, 64]. It has been observed that miR-21 is the most abundant miRNA in macrophages and its deficiency in vivo in mice has been shown to promote atherosclerosis and vascular inflammation [65].

3.1.4 MiR-132

MiR-132 was one of the first miRNAs, together with miR-146a and miR-155, found upregulated after TLR4 engagement by LPS in THP-1 human monocytic cells [20, 21]. These results were confirmed also in vitro in RAW 264.7 murine macrophage-derived cell line and in vivo in murine splenocytes after treatment with LPS [66,

67]. Validated target of miR-132 are IRAK4 [68] and acetylcholinesterase [67, 69], thus indicating a role for this miRNA in regulating the LPS signaling pathway and also in the crosstalk between immune and neuronal systems.

3.2 *MiRNA Downregulated by TLR4 Triggering*

3.2.1 MiR-223

MiR-223 is preferentially expressed in hematopoietic system, having the highest expression in granulocytes [70]. In human monocytes, differentiation into macrophages is accompanied by decreased expression of miRNA-223 [71]. In human monocyte–derived immature dendritic cells stimulated with LPS a slight increase of miR-223 was documented by Ceppi and colleagues [46]. Indeed, miR-223 has been found upregulated in M2-polarized macrophages in comparison with M1-polarized macrophages [72]. Several studies in mouse models have demonstrated that activation of TLR4 by LPS downregulates miR-223 in macrophages [73–75]. MiR-223-deficient macrophages have been shown to be hypersensitive to LPS stimulation; this indicates that miR-233 actively participate in downregulating the proinflammatory response induced via TLR4 [74]. MiR-223 acts by repressing IκB kinase subunit-α (IKK-α) [71, 76], STAT3 [73, 76] and the nucleotide-binding oligomerization domain-like receptor (NLR) inflammasome by targeting the NLR protein 3 (NLRP3) 3′-UTR [77, 78]. These observations indicate that miR-223 represents a fine-tuner of inflammation both directly on TLR4 signaling pathway, both indirectly on cytokine signaling mechanisms and inflammasome activity.

3.2.2 MiR-125b

Tili and colleagues [45] were the first demonstrating that TLR4 stimulation by LPS produces a downregulation of miR-125b both in vitro in RAW264.7 murine macrophages both in vivo. Similar results have been obtained more recently by Murphy AJ et al. [79] in human macrophages and by Jia J et al. [80] in a murine chondrogenic cell line. Conversely, other papers showed an increased miR-125b expression after LPS challenge in human monocytes and mouse RAW264.7 [81, 82]. At present, no data are available to explain this discrepancy; it is likely that the timing for the measure of miR-125b after LPS challenge could play a critical role. Indeed, all the papers agree that miR-125b actively cooperates in controlling excessive proinflammatory response via TNF-α post-transcriptional repression, targeting the 3′-untranslated region of TNF-alpha mRNA. Increased expression of miR-125b was found in endotoxin tolerance, by inhibiting TNF-α translation [83]. Other verified targets of miR-125b were MIP-1α [80] and mitochondrial metabolism through post-transcriptional repression of the BH3-only proapoptotic protein BIK and the mitochondrial fission process 1 protein MTP18 [81].

3.3 MiRNA Targeting TLR4 Expression and/or Signaling Proteins

There are some miRNAs that are not directly induced or downregulated by TLR4 engagement but can influence TLR4 expression and/or translation of signaling proteins involved in TLR4 pathway. MiRNAs of Let-7 family, such as let-7i, let-7e, let-7a, let-7b were shown to repress TLR4 [79, 84–86]. A direct effect of miR-511 on human TLR4 3′-UTR was also demonstrated by Tserel and colleagues in 2011 [87] and, more recently, confirmed by Curtale and colleagues [88]. Members of the MiR-200 family, namely miR-200b and miR-200c, have been demonstrated to target some signaling proteins involved in TLR4 pathway, such as MyD88 and c Jun/MAPK [89, 90].

4 TLR4–miRNA Axis in Nonimmune Cells

Even though TLR4 expression has been firstly described in innate immune cells, in the last decades several reports have documented the expression of this receptor on the cell membrane of several other cellular subsets, including epithelial [91] and endothelial cells [92, 93], cardiac myocytes [94, 95], mesenchymal stromal cells [96], and tumor cells [97–99]. As a consequence of the presence of TLR4, similar mechanisms of TLR4–miRNA activation have been demonstrated in nonimmune cells, directly contributing to the fine regulation of cellular homeostasis. It has been demonstrated that TLR4 activation in mesenchymal stromal cells (MSCs) modulates inflammatory responses by targeting components of the TLR signaling pathway both in MSCs and in cells that interact with MSCs, through the release of extracellular vesicle containing miRNAs [100–102]. In primary lung cancer cells, direct activation of TLR4 by LPS has been shown to enhance tumor outgrowth in vitro and in vivo by inducing miR-21. This pathway results in being particularly relevant in consideration of the direct correlation between TLR4 and miR-21 expressions observed in freshly isolated, untreated primary human lung cancer cells [103]. MiR-145-5p has been recently found to exhibit antitumorigenic activity in vitro and in vivo in melanoma by suppressing TLR4 expression and inactivating NF-κB pathway [104]. In primary human microvascular endothelial cells, overexpression of miR146a by miRNA mimic has been demonstrated to significantly reduce the level of TLR4/NF-κB activation induced by the presence of high glucose concentrations [105].

5 Crosstalk Among Different Cellular Subsets Mediated by miRNAs in Exosomes Released After TLR4 Stimulation

Recent results have clearly demonstrated that miRNAs post-transcriptionally regulate gene expression within the cells in which they are produced but have also a significant impact at distance on different target cells. This effect is due to the release of extracellular microvesicles (e.g., exosomes) transporting miRNAs that can be taken up by recipient cells, causing changes in their function. McDonald and colleagues showed that murine RAW 264.7 macrophages and THP1 monocytic cells stimulated with LPS are characterized by altered expression in their exosomes of some miRNAs [106]. Similar results have been obtained by Ortega FJ et al. [107]. In 2015, Alexander and colleagues demonstrated that endogenous miR-155 and miR-146a are released by primary murine dendritic cells and, following uptake, induce the reprogramming of target dendritic cell response to LPS, respectively enhancing or reducing inflammatory gene expression [29]. The important role played by miRNAs produced and released after TLR4 activation is highlighted by the observation that miRNA-mediated cellular crosstalk is active in the intercellular communications among various cellular subsets and not limited to the immune compartment [106, 107]. In a diabetic rat model, exosomes released from LPS-activated MSCs downregulate inflammation by directing macrophage polarization and favoring regenerative properties through the production of let-7b miRNA targeting TLR4 protein [102]. Macrophages residing in adipose tissue of obese mice have been shown to produce exosomes containing miRNAs, in particular miR-155, which contributes to the inhibition of insulin signaling [108].

Tumor-associated macrophages have been shown to promote breast cancer invasion and metastasis through the release of miRNA-containing exosomes. In this context, a critical role of miR-223 has been observed: In vitro invasiveness of breast cancer cells was significantly decreased when miR-223 expression by macrophages was abolished by an antisense oligonucleotide [109]. On the other hand, tumor cells have been demonstrated to employ exosomes containing miRNAs to block innate immune activation. Pancreatic cancer-derived exosomes containing miR-203 have been demonstrated to inhibit dendritic cell activation, by inhibiting TLR4 expression and cytokine production [110].

6 Conclusions

There is increasing evidence that TLR4 and miRNA crosstalk ensures the fine-tuning of inflammatory response and subsequent healing occurring in tissues after infection or injury. The expression of miRNAs inside the cell, after TLR4 triggering, critically contributes both to the activation and to the shutdown of immune cell response needed for the termination of the inflammatory process. Moreover, it has been found that miRNAs have a role outside the cell, mediating paracrine intercellular communications, thus extending the influence of TLR4 activation at distance through a wider range of mediators than thought before. These interactions involve both immune and nonimmune cells, indicating how these mechanisms represent an important means to control tissue functions. Pharmacological targeting of these pathways with new treatments could be relevant for those pathological conditions in which uncontrolled inflammatory response and/or aberrant tissue healing occurs.

References

1. Molteni M, Gemma S, Rossetti C. The role of toll-like receptor 4 (TLR4) in infectious and non-infectious inflammation. Mediators of Inflamm. 2016;6978936:1–9.
2. Molteni M, Bosi A, Rossetti C. Natural products with toll-like receptor 4 (TLR4) antagonist activity. Int. J. Inflammation. 2018;2859135:1–9.
3. Baltimore D, Boldin MP, O'Connell R, Rao DS, Taganov KD. Micro-RNAs: new regulators of immune cell development and function. Nature Immunol. 2008;9:839–43.
4. Bartel DP. MicroRNAs: genomics, biogenesis, mechanism and function. Cell. 2004;116:281–97.
5. Bartel DP. MicroRNAs: target recognition and regulatory function. Cell. 2009;136:215–33.
6. Hammond SM. An overview of microRNAs. Adv Drug Del Rev. 2015;87:3–14.
7. Alles J, Fehlmann T, Ischer U, et al. An estimate of the total number of true human miRNAs. Nucl Acid Res. 2019;47:3353–64.
8. Catalanotto C, Cogoni C, Zardo G. microRNA in control of gene expression: an overview of nuclear functions. Int J Mol Sci. 2016;17:E1712.
9. Krol J, Loedige I, Filipowicz W. The widespread regulation of microRNA biogenesis, function and decay. Nat Rev Genetics. 2010;11:597–610.
10. O'Brien J, Hayder H, Zared Y, Peng C. Overview of microRNA biogenesis, mechanisms of actions and circulation. Front Endocrinol. 2018;9:402.
11. Denli AM, Tops BB, Plasterk RH, Ketting RF, Hannon GJ. Processing of primary microRNAs by the microprocessor complex. Nature. 2004;432:231–5.
12. Han J, Lee Y, Yeon KH, Kim YK, Jin H, Kim UN. The Drosha-DGCR8 complex in primary microRNA processing. Genes Dev. 2004;18:3016–27.
13. Okada C, Yamashita E, Lee SJ, Shibata S, Katahira J, Nakagawa A, Yoneda Y, Tsukihara T. A high resolution structure of the pre-microRNA nuclear export machinery. Science. 2009;326:1275–9.
14. Kim YK, Kim B, Kim VN. Re-evaluation of the role of DROSHA, export in 5 and DICER in microRNA biogenesis. Proc Natl Acad Sci U S A. 2016;113:E1881–9.
15. Jo MH, Shin S, Jung SR, Kim E, Song JJ, Holing S. Human Argonaute 2 has diverse reaction pathways on target RNAs. Mol Cell. 2015;59:117–24.
16. Huntzinger E, Izaurralde E. Gene silencing by microRNAs: contributions of translational repression and mRNA decay. Nat Rev Genetics. 2011;12:99–110.
17. Liu J, Valencia-Sanchez MA, Hannon G, Parker R. Micro-RNA dependent localization of targeted mRNAs to mammalian P-bodies. Nat Cell Biol. 2005;7:719–23.
18. Filipowicz W, Bhattacharyya SN, Sonenberg N. Mechanisms of post-transcriptional regulation by microRNAs: are the answers in sight? Nat Rev Genetics. 2008;9:102–14.
19. Xiao M, Li J, Li W, et al. microRNAs activate gene transcription epigenetically as an enhancer trigger. RNA Biol. 2017;14:1326–34.

20. Taganov KD, Boldin MP, Chang KJ, Baltimore D. NF-kappaB-dependent induction of microRNA miR-146, an inhibitor targeted to signaling proteins of innate immune responses. Proc Natl Acad Sci U S A. 2006;103:12481–6.
21. Nahid MA, Pauley KM, Satoh M, Chan EKL. miR-146° is critical for endotoxin-induced tolerance. J Biol Chem. 2009;284:34590–9.
22. Brown BD, Gentner B, Cantore A, et al. Endogenous microRNA can be broadly exploited to regulate transgene expression according to tissue lineage and differentiation state. Nat Biotech. 2007;25:1457–67.
23. Molteni M, Bosi A, Saturni V, Rossetti C. MiR-146a induction by cyanobacterial lipopolysaccharide antagonist (CyP) mediates endotoxin cross-tolerance. Sci Rep. 2018;8:11367.
24. Juknat A, Gao F, Coppola G, Vogel Z, Kozela E. miRNA expression profiles and molecular networks in resting and LPS-activated BV-2 microglia. Effects of cannabinoids. PLoS ONE. 2019;14:e0212039.
25. Hou J, Wang P, Lin L, Liu X, Ma F, An H, Wang Z, Cao X. MicroRNA-146a feedback inhibits RIG-I-dependent type I IFN production in macrophages by targeting TRAF6, IRAK1, IRAK2. J Immunol. 2009;183:2150–8.
26. Cavaillon J-M, Adib-Conquy M. Bench-to-bedside review: endotoxin tolerance as a model of leukocyte reprogramming in sepsis. Crit Care. 2006;10:233.
27. Chassin C, Kocur M, Pott J, Duerr CU, Gutle D, Lotz M, Hornef W. miR146a mediates protective innate immune tolerance in the neonate intestine. Cell Host Microbe. 2010;8:358–68.
28. Banerjee S, Meng J, Das S, et al. Morphine-induced exacerbation of sepsis is mediated by tempering endotoxin tolerance through modulation of miR146a. Sci Rep. 2013;3:177.
29. Alexander M, Hu R, Runtsch MC, et al. Exosome-delivered microRNAs modulate inflammatory response to endotoxin. Nat Commun. 2015;6:7321.
30. Boldin MP, Taganov D, Rao DS, et al. miR146a is a significant brave on autoimmunity, myeloproliferation, and cancer in mice. J Exp Med. 2011;208:1189–201.
31. Renzi TA, Rubino M, Gornati L, Garlanda C, locate M, Curtale G. miR-146b mediates endotoxin tolerance in human phagocytes. Mediators Inflamm. 2015;145305
32. Curtale G, Mirolo M, Renzi TA, Rossato M, Bazzoni F, Locati M. Negative regulation of toll-like receptor 4 signaling by IL-10-dependent microRNA-146b. Proc. Natl Acad Sci USA. 2013;110:11499–504.
33. O'Connell RM, Kahn D, Gibson WS, et al. MicroRNA-155 promotes autoimmune inflammation by enhancing inflammatory T cell development. Immunity. 2010;33:607–19.
34. Thai TH, Calado DP, Casola S, et al. Regulation of the germinal center response by microRNA-155. Science. 2007;316:604–8.
35. Vigorito E, Kohlhaas S, Lu D, Leyland R. MiR155: an ancient regulator of immune system. Imm Rev. 2013;253:146–57.
36. O'Connell RM, Taganov KD, Boldin MP, Cheng G, Baltimore D. MicroRNA-155 is induced during the macrophage inflammatory response. Proc Natl Acad Sci U S A. 2007;104:1604–9.
37. Cardoso AL, Guedes JR, Pereira de Almeida L, Pedroso de Lima MC. MiR155 modulates microglia-mediated immune response by downregulating SOCS-1 and promoting cytokine and nitric oxide production. Immunology. 2012;135:73–88.
38. Mann M, Mehta A, Zhao JL, et al. An NF-κB-microRNA regulatory network tunes macrophage inflammatory responses. Nat Commun. 2017;8:851.
39. Doxaki C, Kampranis SC, Eliopoulos AG, Spilianakis C, Tsatsanis C. Coordinated regulation of miR-155 and miR-146a genes during induction of endotoxin tolerance in macrophages. J Immunol. 2015;195:5750–61.
40. Elton TS, Selemon H, Elton SM, Parinandi NL. Regulation of the miR-155 host gene in physiological and pathological processes. Gene. 2013;532:1–12.
41. O'Connell RM, Chaudhuri AA, Rao DS, Baltimore D. Inositol phosphate SHIP1 is a primary target of miR155. Proc Natl Acad Sci U S A. 2009;106:7113–8.
42. An H, Xu H, Zhang M, Zhou J, Feng T, Qian C, Qi C, Cao X. Src homology 2 domain-containing inositol-5-phosphatase 1 (SHIP1) negatively regulates TLR4-mediated LPS

response primarily through a phosphatase activity- and PI-3K-independent mechanism. Blood. 2005;105:4685–92.
43. Wang P, HOu J, Lin L, et al. Inducible miR-155 feedback promotes type 1 IFN signaling in antiviral innate immunity by targeting suppressor of cytokine signaling 1. J Immunol. 2010;185:6226–33.
44. Nazari-Jahantigh M, Wei Y, Noels H, et al. MicroRNA-155 promotes atherosclerosis by repressing BCL6 in macrophages. J Clin Invest. 2012;122:4190–202.
45. Tili E, Michaille JJ, Cimino A, et al. Modulation of miR-155 and miR-125b levels following lipopolysaccharide/TNFα stimulation and their possible roles in regulating the response to endotoxin shock. J Immunol. 2007;179:5082–9.
46. Ceppi M, Pereira PM, Dunand-Sauthier I, Barras E, Reith W, Santos MA, Pierre P. MicroRNA-155 modulates the interleukin-1 signaling pathway in activated human monocyte-derived dendritic cells. Proc Natl Acad Sci U S A. 2009;106:2735–40.
47. Bala S, Marcos M, Kodys K, Csak T, Catalano D, Mandrekar P, Szabo G. Upregulation of micro RNA-155 in macrophages contributes to increased tumor necrosis factor α (TNFα) production via increased mRNA half-life in alcoholic liver disease. J Biol Chem. 2011;286:1436–44.
48. O'Connell RM, Rao DS, Chaudhuri AA, Boldin MP, Taganov KD, Nicoll J, Paquette RL, Baltimore D. Sustained expression of microRNA 155 in hematopoietic stem cells causes a myeloproliferative disorder. J Exp Med. 2008;205:585–94.
49. Cai X, Yin Y, Li N, Zhu D, Zhang J, Zhang CY, Zen K. Re-polarization of tumor-associated macrophages to pro-inflammatory M1 macrophages by microRNA-155. J Mol Cell Biol. 2012;4:341–3.
50. Lin EF, Millar DG, Dissanayake D, Savage JC, Grimshaw NK, Kerr WG, Ohashi PS. miR-155 upregulation in dendritic cells is sufficient to break tolerance in vivo by negatively regulating SHIP1. J Immunol. 2015;195:4632–40.
51. Rodriguez A, Vigorito E, Clare S, et al. Requirement of bic/microRNA-155 for normal immune function. Science. 2007;316:608–11.
52. Lind EF, Elford AR, Ohashi PS. MicroRNA 155 is required for optimal CD8+ T cell responses to acute and intracellular bacterial challenge. J Immunol. 2013;190:1210–6.
53. Ji Y, Fioravanti J, Zhu W, et al. miR-155 harnesses phf19 to potentiate cancer immunotherapy through epigenetic reprogramming of CD8+ T cell fate. Nat Comm. 2019;10:2157.
54. Alivernini S, Gremese E, McSharry C, tolusso B, ferraccioli G, McInnes IB, Kurowska-Stolarska M. MicroRNA-155- at the critical interface of innate and adaptive immunity in arthritis. Front Immune. 2018;8:1932.
55. Kurowska-Stolarska M, Alivernini S, Ballantine LE, et al. MicroRNA-155 as a pro-inflammatory regulator in clinical and experimental arthritis. Proc Natl Acad Sci U S A. 2011;108:11193–8.
56. Sheedy FJ, Palsson-McDermott E, Hennessy EJ, et al. Negative regulation of TLR4 via targeting of the proinflammatory tumor suppressor PDCD4 by the microRNA miR-21. Nat Immunol. 2010;11:141–7.
57. Sheedy FJ. Turning 21: induction of miR-21 as a key switch in the inflammatory response. Front Immunol. 2015;6:19.
58. Gunzl P, Schabbauer G. Recent advances in the genetic analysis of PTEN and PI3K innate immune properties. Immunobiology. 2008;213:759–65.
59. Das A, Ganesh K, Khanna S, Sen CK, Roy S. Engulfment of apoptotic cells by macrophages: a role for microRNA-21 in the resolution of wound inflammation. J Immunol. 2014;192:1120–9.
60. Barnett RE, Conklin DJ, Ryan L, et al. Anti-inflammatory effects of miR-21 in the macrophage response to peritonitis. J Leuk Biol. 2016;99:361–71.
61. Xi J, Huang Q, Wang L, et al. miR-21 depletion in macrophages promotes tumoricidal polarization and enhances PD-1 immunotherapy. Oncogene. 2018;37:3151–65.

62. Iliopoulos D, Jaeger S, Hirsch H, Bulyk L, Struhl K. STAT3 activation of miR-21 and miR-181b-1 via PTEN and CYLD are part of the epigenetic switch linking inflammation to cancer. Mol Cell. 2010;39:493–506.
63. Lu TX, Munitz A, Rothenberg ME. miR-21 is upregulated in allergic airway inflammation and regulates IL-12 p35 expression. J Immunol. 2009;182:4994–5002.
64. Lu TX, Hartner J, Lim E-J, et al. miRNA-21 limits in vivo immune response-mediated activation of the IL12/IFNγ pathway, Th1 polarization and the severity of delayed-type hypersensitivity. J Immunol. 2011;187:3362–73.
65. Canfran-Duque A, Rotland N, Zhang X, et al. Macrophage deficiency of miR-21 promotes apoptosis, plaque necrosis, and vascular inflammation during atherogenesis. EMBO Mol Med. 2017;9:1244–62.
66. Kanaan Z, Barnett R, Gardner S, Keskey B, Druen D, Billeter A, Cheadle WG. Differential microRNA (miRNA) expression could explain microbial tolerance in a novel chronic peritonitis model. Innate Immun. 2013;19:203–12.
67. Shaked I, Meerson A, Wolf Y, Avni R, Greenberg D, Gilboa-Geffen A, Soreq H. MicroRNA-132 potentiates cholinergic anti-inflammatory signaling by targeting acetylcholinesterase. Immunity. 2009;31:965–73.
68. Nahid MA, Yao B, Dominguez-Gutierrez PR, Kesavalu L, Satoh M, Chan EK. Regulation of TLR2-mediated tolerance and cross-tolerance through IRAK4 modulation by miR-132 and miR-212. J Immunol. 2013;190:1250–63.
69. Liu F, LiY JR, et al. miR-132 inhibits lipopolysaccharide-induced inflammation in alveolar macrophages by the cholinergic anti-inflammatory pathway. Exp Lung Res. 2015;41:261–9.
70. Ye D, Zhang T, Lou G, liu Y. Role of miR-223 in the pathophysiology of liver disease. Exp Mol Med. 2018;50:128.
71. Li T, Morgan MJ, Choksi S, Zhang Y, Kim Y-S, Liu Z-G. microRNAs modulate the noncanonical transcription factor NF-κB pathway by regulating expression of the kinase IKKα during macrophage differentiation. Nat Immunol. 2010;11:799–805.
72. Liu G, Abraham E. microRNAs in immune response and macrophage polarization. Arterioscler Thromb Vasc Biol. 2013;33:170–7.
73. Chen Q, Wang H, Liu Y, Song Y, Lai L, Han Q, Cao X, Wang Q. Inducible microRNA 223 downregulation promotes TLR-triggered IL-6 and IL1β production in macrophages by targeting STAT3. PLoS One. 2012;7:e42971.
74. Zhuang G, Meng C, Guo X, et al. A novel regulator of macrophage activation: miR-223 in obesity-associated adipose tissue inflammation. Circulation. 2012;12:2892–903.
75. Wang J, Bai X, Song Q, Fan F, Hu Z, Cheng G, Zhang Y. miR-223 inhibits lipid deposition and inflammation by suppressing toll-like receptor 4 signaling in macrophages. Int J Mol Sci. 2015;16:24965–82.
76. Wu J, Niu P, Zhao Y, et al. Impact of miR-223-3p and miR-2909 on inflammatory factors IL6, IL1β, and TNFα, and the TLR4/TLR2/NFκB/STAT3 signaling pathway induced by lipopolysaccharide in human adipose stem cells. PLoS One. 2019;14:e0212063.
77. Bauernfeind F, Rieger A, Schildberg FA, Knolle PA, Schmid-Burgk JL, Hornung V. NLRP3 inflammasome activity is negatively controlled by miR-223. J Immunol. 2012;189:4175–81.
78. Haneklaus M, Gerlic M, Kurowska-Stolarska M, et al. Cutting edge: miR223 and EBV miR-BART15 regulate the NLRP3 inflammasome and IL-1β production. J Immunol. 2012;189:3795–9.
79. Murphy AJ, Guyre PM, Pioli PA. Estradiol suppresses NF-κB activation through coordinated regulation of let-7a and miR-125b in primary human macrophages. J Immunol. 2010;184:5029–37.
80. Jia J, Wang J, Zhang J, Cui M, Sun X, Li Q, Zhao B. miR125b inhibits LPS-induced inflammatory injury via targeting MIP-1α in chondrogenic cell ATDC5. Cell Physiol Biochem. 2018;45:2305–16.
81. Duroux-Richard I, Roubert C, Ammari M, et al. miR-125b controls monocyte adaptation to inflammation through mitochondrial metabolism and dynamics. Blood. 2016;128:3125–36.

82. Guan Y, Yao H, Wang J, Sun K, Cao L, Wang Y. NF-κB-DICER-miRs axis regulates TNFα expression in responses to endotoxin stress. Int J Biol Sci. 2015;11:1257–68.
83. El Gazzar M, McCall CE. microRNAs distinguish translational from transcriptional silencing during endotoxin tolerance. J Biol Chem. 2010;285:20940–51.
84. Androulidaki A, Iliopoulos D, Arranza A, et al. The kinase akt1 controls macrophage response to lipopolysaccharide by regulating microRNAs. Immunity. 2009;31:220–31.
85. Chen XM, Splinter PL, O'Hara SP, LaRusso NF. A cellular micro-RNA, let-7i, regulates toll-like receptor 4 expression and contributes to cholangiocyte immune responses against Cryptosporidium parvum infection. J Biol Chem. 2007;282:28929–38.
86. Teng GG, Wang WH, Dai Y, Wang SJ, Chu YX, Li J. Let-7b is involved in the inflammation and immune responses associated with helicobacter pylori infection by targeting toll-like receptor 4. PLoS One. 2013;8:e56709.
87. Tserel L, Runnel T, Kisand K, et al. MicroRNA expression profiles of human blood monocyte-derived dendritic cells and macrophages reveals miR-511 as putative positive regulator of toll-like receptor 4. J Biol Chem. 2011;286:26487–95.
88. Curtale G, Renzi TA, Drufuca l, Rubino M, Locati M. Glucocorticoids downregulate TLR4 signaling activity via its direct targeting by miR-511-5p. Eur J Immunol. 2017;47:20870–2089.
89. Wendlandt EB, Graff JW, Gioannini TL, McCaffrey AP, Wilson ME. The role of microRNAs miR-200b and miR-200c in TLR4 signaling and NF-κB activation. Innate Immun. 2012;18:846–55.
90. Jadhav SP, Kamath SP, Choolani M, Lu J, Dheen ST. MicroRNA-200b modulates microglia-mediated neuroinflammation via the cjun/MPK pathway. J Neurochem. 2014;130:388–401.
91. Backhed F, Hornef M. Toll-like receptor 4-mediated signaling by epithelial surfaces: necessity or threat? Micr Infect. 2003;5:951–9.
92. Zeuke S, Ulmer AJ, Kusumoto S, Katus HA, Heine H. TLR4-mediated inflammatory activation of human coronary artery endothelial cells by LPS. Cardiovasc Res. 2002;56:126–34.
93. Mudaliar H, Pollock C, Ma J, Wu H, Chadban S, Panchapakesan U. The role of TLR2 and 4-mediated inflammatory pathways in endothelial cells exposed to high glucose. PLoS One. 2014;9:e108844.
94. Frantz S, Kobzik L, Kim Y-D, Fukazawa R, Medzhitov R, Lee RT, Kelly RA. Toll4 (TLR4) expression in cardiac myocytes in normal and failing myocardium. J Clin Invest. 1999;104:271–80.
95. Liu L, Wang Y, Cao ZY, et al. Up-regulated TLR4 in cardiomyocytes exacerbates heart failure after long-term myocardial infarction. J Cell Mol Med. 2015;19:2728–40.
96. Raicevic G, Rouas R, Najar M, et al. Inflammation modifies the pattern and the function of toll-like receptors expressed by human mesenchymal stromal cells. Human Immunol. 2010;71:235–44.
97. Molteni M, Marabella D, Orlandi C, Rossetti C. Melanoma cell lines are responsive in vitro to lipopolysaccharide and express TLR-4. Cancer Lett. 2006;235:75–83.
98. Chen X, Zhao F, Zhang H, Zhu Y, Wu K, Tan G. Significance of TLR4/MyD88 expression in breast cancer. Int J Clin Exp Pathol. 2015;8:7034–9.
99. Pinto A, Morello S, Sorrentino R. Lung cancer and toll-like receptors. Cancer Immunol Immunother. 2011;60:1211–20.
100. Abdi J, Rashedi I, Keating A. Concise review: TLR pathway-miRNA interplay in mesenchymal stromal cells: regulatory roles and therapeutic directions. Stem Cells. 2018;36:1655–62.
101. Liu G-Y, Liu Y, Lu Y, et al. Short-term memory of danger signals or environmental stimuli in mesenchymal stem cells: implications for therapeutic potential. Cell & Moll Immunol. 2016;13:369–78.
102. Ti D, Hao H, Tong C, et al. LPS-preconditioned mesenchymal stromal cells modify macrophage polarization for resolution of chronic inflammation via exosome-shuttles let-7b. J Transl Med. 2015;13:308.

103. Zhang X, Wang C, Shan S, Liu X, Jiang Z, Ren T. TLR4/ROS/miRNA-21 pathway underlies lipopolysaccharide instructed primary tumor outgrowth in lung cancer patients. Oncotarget. 2016;7:42172–82.
104. Jin C, Wang A, Liu L, Wang G, Li G, Han Z. miR145-5p inhibits tumor occurrence and metastasis through the NF-κB signaling pathway by targeting TLR4 in malignant melanoma. J Cell Biochem. 2019;20:11115–26.
105. Ye E-A, Steile JJ. miR146a attenuates inflammatory pathway mediated by TLR4/NF-κB and TNF-α to protect primary human retinal microvascular endothelial cells grown with high glucose. Mediators Inflamm. 2016;3958453
106. McDonald MK, Tian Y, Qureshi A, et al. Functional significance of macrophage-derived exosomes in inflammation and pain. Pain. 2014;155:1527–39.
107. Ortega FJ, Moreno M, Mercader JM, Moreno-Navarrete JM, Fuentes-Batllevell N, Sabater M, Ricart W, Fernandez-Real JM. Inflammation triggers specific microRNA profiles in human adipocytes and macrophages and in their supernatants. Clin Epigenetics. 2015;7:49.
108. Ying W, Riopel M, Bandyopadhyay G, et al. Adipose tissue macrophage-derived exosomal miRNAs can modulate *in vivo* and *in vitro* insulin sensitivity. Cell. 2017;171:372–84.
109. Yang M, Chen J, Su F, et al. Microvesicles secreted by macrophages shuttle invasion-potentiating microRNAs into breast cancer cells. Mol Cancer. 2011;10:117.
110. Zhou M, Chen J, Zhou L, Chen W, Ding G, Cao L. Pancreatic cancer derived exosomes regulate the expression of TLR4 in dendritic cells via miR-223. Cell Immunol. 2014;292:65–9.

Metabolic Reprogramming of Myeloid Cells Upon TLR4 Stimulation

Laure Perrin-Cocon, Anne Aublin-Gex, and Vincent Lotteau

Abstract Cell metabolism sustains the generation of energy, the modulation of cell signaling, and the production of molecular blocks for cell maintenance and/or proliferation. Immune cells not only require energy to survive but also need to adapt their metabolism in response to activation signals in order to differentiate and assume their immune functions. Numerous studies now indicate that the function of innate immune cells is closely related to their metabolic status. Hence, profound metabolic reprogramming occurs upon stimulation of pathogen recognition receptors (PRR), especially in myeloid cells. This review mainly focuses on the crosstalk between toll-like receptor 4 (TLR4) signaling and metabolic regulations occurring in macrophages and dendritic cells (DCs). Multiple studies showed that TLR4 stimulation of DCs or macrophages results in increased glycolytic activity, an essential process to support their proinflammatory functions. However, the molecular mechanisms involved have been only partially discovered and differ according to cell types and species. This chapter gives a knowledge overview of the molecular mechanisms involved in the modulation of central carbon metabolism, from glycolysis, tricarboxylic acid (TCA) cycle, oxidative phosphorylation, to lipid metabolism, upon TLR4 signaling in macrophages and DCs.

Keywords Innate immunity · TLR4 signaling · Cell metabolism · Macrophage · Dendritic cell · Glycolysis · TCA cycle · Oxidative phosphorylation · Lipid metabolism

Cell metabolism sustains the generation of energy, the modulation of cell signaling and the production of molecular blocks for cell maintenance and/or proliferation. An increasing amount of evidence points to a strong intertwined relationship between the metabolic state and the effective function of immune cells, leading to the emergence of the immunometabolism field [1]. It was first reported in the 1960s

L. Perrin-Cocon (✉) · A. Aublin-Gex · V. Lotteau
CIRI, Centre International de Recherche en Infectiologie, Univ Lyon, Inserm, U1111, Université Claude Bernard Lyon 1, CNRS, UMR5308, ENS de Lyon, Lyon, France
e-mail: laure.perrin@inserm.fr; anne.aublin-gex@inserm.fr; vincent.lotteau@inserm.fr

C. Rossetti, F. Peri (eds.), *The Role of Toll-Like Receptor 4 in Infectious and Non Infectious Inflammation*, Progress in Inflammation Research 87,
https://doi.org/10.1007/978-3-030-56319-6_11

that immune cells activation correlated to specific metabolic properties and that blocking metabolic pathways could lead to functional changes [2, 3]. Indeed, immune cells not only require energy to survive but also need to adapt their metabolism in response to activation signals in order to differentiate and assume their immune functions. The type and fate of nutrients used by immune cells differ according to their functional requirements. Therefore, profound metabolic reprogramming occurs in myeloid cells upon stimulation of pathogen recognition receptors (PRR) and certain metabolites can modulate the function of these immune cells [4, 5]. This chapter will mainly focus on the studies analyzing the crosstalk between toll-like receptor 4 (TLR4) signaling and metabolic regulations occurring in macrophages and dendritic cells (DCs).

1 TLR4 Stimulation Triggers a Warburg-Like Shift of Cell Metabolism

Glycolysis converts glucose to pyruvate by a series of enzymatic reactions, generating 2 moles of ATP per mole of glucose. Three rate-limiting enzymes are controlling the glycolytic flux (hexokinase, phosphofructokinase, and pyruvate kinase) (Fig. 1). The hexokinase (HK) converts glucose to glucose-6-phosphate and controls the entry of glucose-derived carbon into the catabolic pathway. There are four isoenzymes of HK (HK-I, II, III, and IV) encoded by four different genes (HK1, 2, 3, and 4). HK1 and HK3 have a large spectrum of tissue expression. HK4 is expressed in the liver and pancreas, whereas HK2 is overexpressed in many cancer cells and is induced in innate immune cells such as MoDCs upon TLR4 stimulation [6]. The pyruvate produced by glycolysis can be either converted to lactate that is excreted, or enter the mitochondria to be converted into acetyl-CoA or oxaloacetate to fuel the tricarboxylic acid (TCA) cycle (Fig. 1). Under aerobic conditions, this metabolic pathway is coupled to the respiratory chain, allowing electron transport and oxidative phosphorylation (OXPHOS), generating high amounts of ATP (36 moles / mole of glucose) (Fig. 2). During hypoxia, OXPHOS is reduced and glycolysis is increased to face energetic needs [7]. However, activation of glycolysis can also occur under aerobic conditions, and Otto Warburg first discovered that even when oxygen is available tumor cells have a high rate of glycolysis with reduced mitochondrial activity, most pyruvate being converted to lactate [8]. TLR4 stimulation by lipopolysaccharide (LPS) results in a Warburg-like shift of cell metabolism, enhancing the glycolytic activity of murine macrophages and DCs and reducing the mitochondrial catabolic pathways, despite the abundance of oxygen [5, 9, 10]. Inhibition of hexokinase, the first enzyme of the glycolysis pathway by 2-deoxyglucose (2-DG), strongly impedes the entire activation process, reducing the secretion cytokines, motility properties, and the expression of characteristic phenotypic markers of mature DCs [9–12]. Glycogen metabolism supports early glycolytic reprogramming required for DC immune responses [13]. Thus, upon TLR4 engagement, cells adapt their metabolism to accommodate altered functional outputs where glucose remains a source of energy while becoming a source of

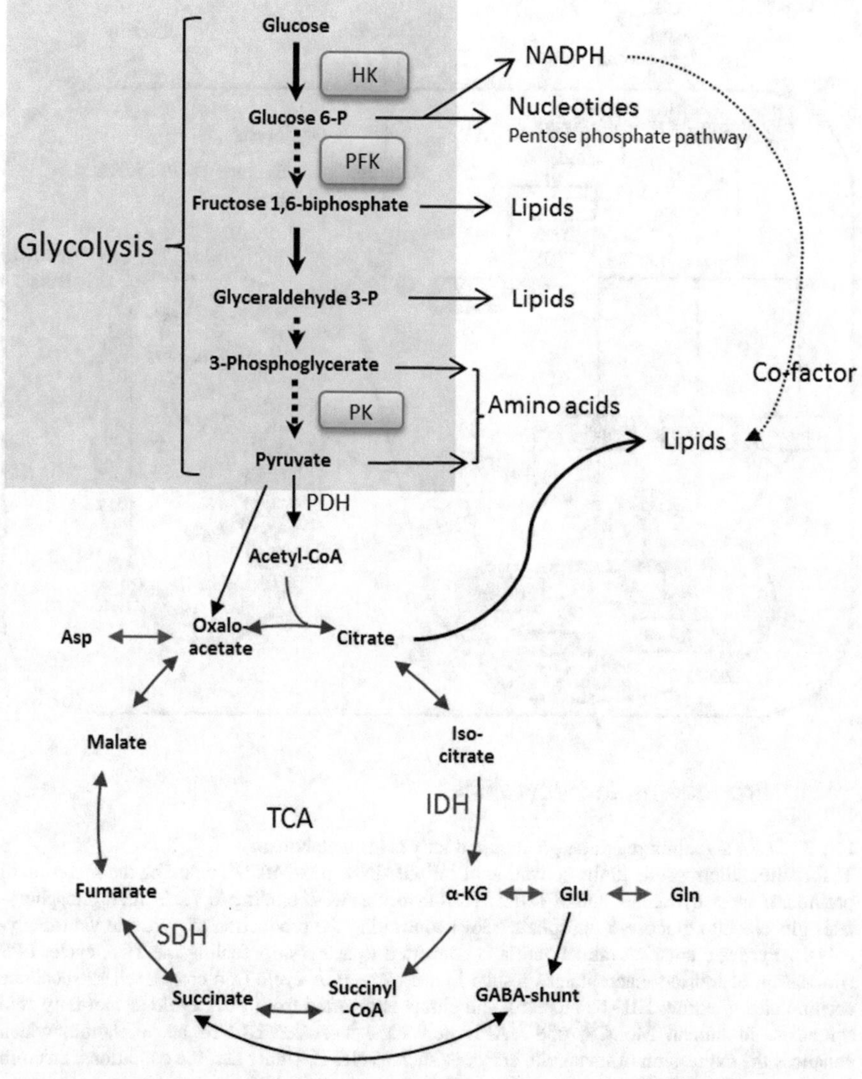

Fig. 1 Glucose catabolic pathway

Glucose is first processed by glycolysis which comprises a series of enzymatic reactions involving three rate-limiting enzymes in blue, generating pyruvate. After entry into the mitochondria, pyruvate is converted into acetyl-CoA or oxaloacetate which combine to generate citrate in the TCA cycle. Glucose metabolism fuels essential anabolic pathways for biosynthesis of nucleotides, lipids, and amino acids. TCA replenishment can be performed using glutamine (Gln) and aspartate (Asp)

carbon for new biosynthetic purposes. Indeed, the increased glycolytic flux results in the accumulation of intermediary metabolites of the pathway that are precursors for the biosynthesis of pentose phosphates, lipids, and amino acids [4] (Fig. 1). However metabolic adaptation depends on the cell type and species.

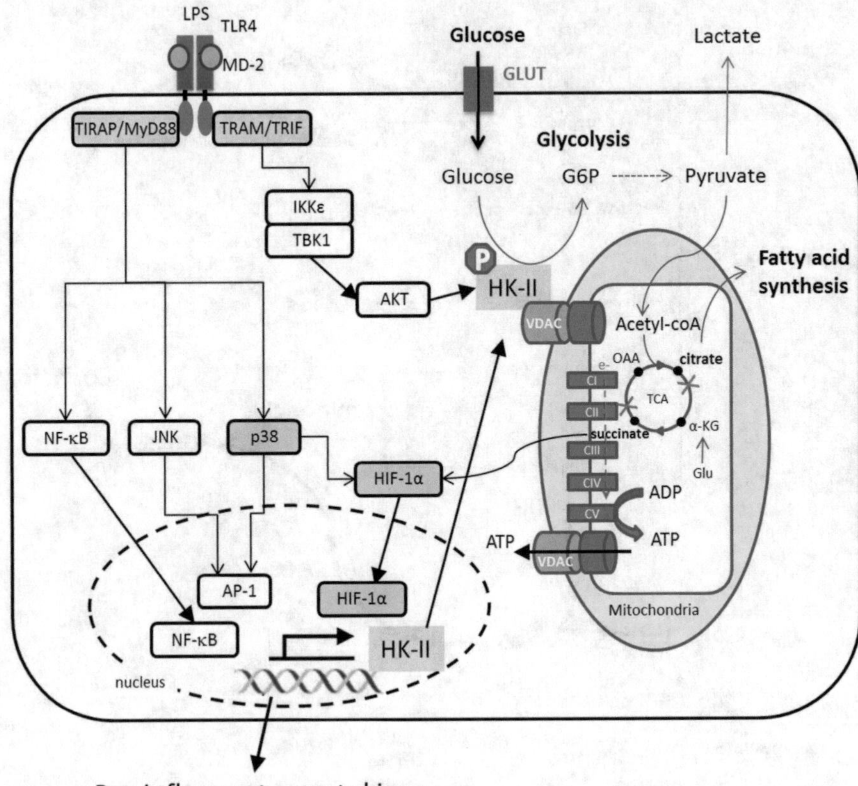

Fig. 2 TLR4 signaling regulates glucose and fatty acids metabolism
TLR4 stimulation results in the activation of NF-κB, JNK, p38-MAPK, inducing the secretion of proinflammatory cytokines, and of TBK1/AKT, favoring HK-II binding to VDAC. HK phosphorylates glucose into glucose-6-phosphate (G6P) controlling the production of pyruvate via the glycolysis. Pyruvate entering mitochondria is converted to acetyl-coA fueling the TCA cycle. LPS stimulation of murine macrophages results in a broken TCA cycle (red cross), where succinate accumulates, favoring HIF-1α increase, and citrate is diverted from TCA cycle to fuel fatty acid synthesis. In human MoDCs, p38-MAPK activation increases HIF-1α accumulation, which enhances the expression of metabolic enzymes such as HK-II. Under aerobic conditions, electron (e-) transport through the respiratory chain (CI to CV) generates ATP

2 Different Metabolic Reprogramming According to Cell Type, Species, and Origin

Multiple studies showed that TLR4 stimulation of DCs or macrophages results in increased glycolytic activity, an essential process to support their proinflammatory functions [5, 6, 9–11, 14–17]. However, human monocytes treated with LPS undergo little modulation of glycolytic activity compared to human monocyte-derived DCs

[11]. Bone marrow–derived macrophages (BM-DMs) respond to LPS with the typical Warburg effect whereas the response of peritoneal macrophages is characterized by the induction of both glycolysis and OXPHOS [18]. Even macrophages from the upper and lower human respiratory tract are metabolically distinct. Macrophages from the upper tract largely rely on glycolysis whereas bronchoalveolar macrophages depend more on mitochondrial respiration [19]. Moreover LPS-activated macrophages exhibit differential metabolic behavior whether cells are differentiated from human monocytes or from mouse bone marrow [16], highlighting differences linked to both species and origin of cells. Macrophages can be classified according to their functional polarization, ranging from classic M1 proinflammatory to alternate M2 anti-inflammatory macrophages and this polarization process has been associated to metabolic reprogramming. A shift to aerobic glycolysis was described to be associated to M1 phenotype of murine inflammatory macrophages, whereas glycolytic activity was unchanged in alternate activated M2 macrophages, which mainly rely on fatty acid oxidation (FAO) [4, 20]. Tolerogenic DCs show a metabolic signature characterized by high glycolytic capacity and high mitochondrial activity fueled by FAO [21].

Reprogramming of carbohydrate metabolism is timely coordinated to the process of macrophage or DC activation [22]. In murine bone marrow-derived macrophages or DCs (BM-DMs or BM-DCs), TLR4 triggering by LPS first induces signaling events engaging the transcriptional response, and leads within the first hour to an immediate increase of the glycolytic flux, fueling the TCA and pentose–phosphate pathways (PPP) [23, 10]. These results suggested that the flux of glucose into the PPP had a crucial role in DC activation by supporting fatty acid (FA) synthesis through the generation of NADPH [10]. Relocalization of HK-II to the mitochondria, where it binds to the voltage-dependent anion channel 1 (VDAC1), is occurring within the first hour of stimulation and is thought to enhance its ATP-dependent activity [10]. Both HK-I and HK-II can bind VDAC in the mitochondrial–endoplasmic reticulum junctions. Binding of HK-I to VDAC promoted respiration in memory CD8+ T cells [24]. After the initial phase, LPS response is amplified by the transcriptional response, resulting after 24 h in increased aerobic glycolysis, reduced mitochondrial respiration, and reconfiguration of the TCA cycle, promoting FA metabolism [22]. LPS stimulation of murine macrophages dampens the activity of the TCA cycle and results in accumulation of some TCA intermediates especially succinate due to succinate dehydrogenase (SDH) inhibition (Fig. 2) [25, 26]. Succinate is an inflammatory signal, favoring stabilization of the transcription factor HIF-1α, thus inducing the secretion of the key inflammatory cytokine IL-1β [25] and the expression of glycolytic enzymes [27]. Glutamine-dependent replenishment of the TCA cycle, providing α-ketoglutarate (α-KG) by the glutamine dehydrogenase or succinate via the γ-amino butyric acid (GABA) shunt, also participates in the regulation of metabolites accumulation (Fig. 1). In macrophages stimulated by LPS, succinate oxidation by SDH also provides electrons allowing the production of ROS from complex I of the respiratory chain by reverse electron transport [26].

Decreased expression of isocitrate dehydrogenase (IDH) participates in TCA reconfiguration in murine M1 macrophages [28], favoring citrate exit from the TCA cycle. Citrate which is exported from the mitochondria to the cytosol by a specific transporter is the main source of carbon for FA synthesis (Fig. 2). In the cytosol, citrate is cleaved into oxaloacetate and acetyl-CoA which is used for FA elongation in Wakil helix. Oxaloacetate is converted to malate and then pyruvate, generating NADPH, an important cofactor for FA synthesis and other anabolic processes. Expression of the mitochondrial citrate carrier is highly induced by LPS in macrophages [29]. In BM-DCs stimulated by LPS for 1 h, citrate is preferentially extracted from the TCA to fuel FA elongation in the cytosol [10]. Thus, citrate availability drives the synthesis of FAs that are needed for ER/Golgi expansion to support cytokine secretion by macrophages and DCs [5, 30, 31]. This is consistent with a study showing that immunogenicity of DCs in the liver correlates with a high lipid intracellular content [32]. Citrate export also plays an important role in the generation of prostaglandins derived from arachidonic FA [29].

3 Metabolic Reprogramming and Trained Immunity

After a first encounter of a pathogen, the innate immune system can retain an innate memory, leading to increased responsiveness upon secondary stimulation. This nonspecific innate immune memory is named trained immunity and is orchestrated by epigenetic reprogramming, eventually protecting the host against secondary infections [33, 34]. In some cases, trained immunity may result in maladaptive states leading to hyperinflammation or immune paralysis, as for example in sepsis. Trained immunity has been historically investigated in vitro using monocytes exposed to LPS or β-glucan (from *C. Albicans*), before restimulation 5 days later to obtain macrophage-like cells. LPS-training of monocytes is associated with silencing of genes coding for inflammatory cytokines, rendering cells refractory to further LPS stimulation (LPS tolerance) and with priming of other genes involved in antimicrobial defense [35]. LPS and β-glucan induce trained immunity through a MAPK-dependent pathway that phosphorylates the transcription factor ATF7, subsequently reducing the repressive histone mark H3K9me2 [36]. These changes lead to modulation of cytokine production and modifications in the metabolic state of the cell [37]. A shift from OXPHOS to aerobic glycolysis triggered by an Akt/mTOR/HIF-1α-dependent pathway has been reported to be essential for trained immunity induced by β-glucan [38]. NAD^+ -dependent Sirtuin histone deacetylases can also contribute to epigenetic modulations of gene expression, mediating a switch from glucose metabolism to fatty acid oxidation upon LPS stimulation of macrophages [39].

Various metabolites are cofactors for epigenetic enzymes, resulting in different trained immunity programs. Important links between altered metabolites profile and epigenetic changes have been made in trained monocytes stimulated by β-glucan and subsequently treated with LPS. Accumulation of fumarate, due to glutamine replenishment of the TCA cycle, induces HIF-1α activation resulting in inhibition of KDM5 histone demethylase and thus increases in the H3K4me3 mark on TNFα gene, enhancing TNFα production [40]. Moreover, acetyl-CoA is required for histone acetylation [33]. As previously mentioned, cytosolic acetyl-CoA mainly originates from mitochondrial efflux of citrate, which is split into oxaloacetate and acetyl-CoA. This later is used for acetylation of proteins both in cytosol and nucleus at their N-terminus and at lysine residues. The relative ratio of TCA cycle intermediates α-ketoglutarate and succinate modulates the activity Fe(II)/α-ketoglutarate-dependent dioxygenases such as prolyl hydroxylases involved in HIF-1α degradation (cf. § 4.1) and those involved in histone and DNA methylation [41]. Succinate and the closely related metabolite fumarate are inhibitors of the JMJ family of lysine demethylases and the TET family of methyl-cytosine hydroxylases [34].

4 Different Molecular Mechanisms Are Involved in Metabolic Reprogramming

4.1 Mechanisms Controlling Glycolytic Reprogramming

The molecular mechanisms controlling glycolytic reprogramming upon TLR4 stimulation that have been partially uncovered differ according to cell types. In murine BM-DCs, activation of TBK1-IKKε and AKT kinases was found to control the early increase of glycolysis by favoring mitochondrial translocation of HK-II, fueling the TCA cycle and FA synthesis [10] (Fig. 2). In these DCs, a late increase in glycolytic metabolism was proposed to be a survival mechanism to maintain ATP production despite the inhibition of OXPHOS by NO, which is produced by the inducible NO synthase (iNOS) [14].

Accumulation of HIF-1α, which is a master transcriptional regulator of glycolytic enzymes, in normoxic conditions has been observed in several cell types upon LPS stimulation [6, 7, 42, 43]. In normoxia, HIF-1α degradation is regulated by hydroxylation of proline and asparagine residues by prolyl-hydroxylase domain enzymes (PHDs). The interaction of the von Hippel–Lindau (VHL) factor with these hydroxylated residues, recruits an E3 ubiquitin ligase that targets HIF-1α to the proteasome for degradation [44]. Hypoxia induces HIF-1α accumulation via PHD inhibition due to the lack of oxygen, its co-substrate [42]. Several molecular mechanisms may result in HIF-1α accumulation upon LPS stimulation, including activation of the mammalian target of rapamycin (mTOR) complex 1. The AKT–mTOR–HIF-1α pathway was found to upregulate glycolysis in murine monocytes [38]. In murine BM-DCs, sequestration of iron, a PHD cofactor, stabilizes HIF-1α upon LPS stimulation [42]. mTOR-dependent HIF-1α activity promotes iNOS expression in LPS-activated murine DCs, regulating late sustained glycolytic reprogramming [45]. In these cells expressing iNOS (limited to murine DCs and minor subsets of human DCs), glucose-sensing by mTORC1/HIF-1α/iNOS circuit impacts both DC metabolism and function [46]. In murine macrophages, LPS-induced HIF-1α accumulation seems to require NF-κB- and ERK-dependent transcriptional events [47, 48]. In other studies, ROS production and succinate accumulation upon LPS stimulation inhibit PHD activity thus increasing HIF-1α stability [25, 49]. In human MoDCs, our results point to a different molecular mechanism since TLR4 activation of glycolysis relied on the activation of p38-MAPK controlling HIF-1α accumulation whose transcriptional activity increased HK2 expression [6]. The p38-MAPK/HIF-1α/HK2 axis was important for cytokine secretion in LPS-stimulated MoDCs [6], whereas the expression of HLA-DR, CD86, and CD40 molecules were unaltered. Interestingly, this pathway did not appear to be involved in the activation of glycolysis triggered by TLR1/2 or TLR2/6 stimulation [6].

Although the molecular mechanisms triggered by TLR4 to stimulate glycolysis are different in human MoDCs and mouse BM-DCs, they both involve HK-II protein modulation resulting in enhanced HK activity [6, 10]. The contribution of other nonenzymatic functions of HK-II should also be analyzed. Indeed, overexpression and mitochondrial association of HK-II confer protection against apoptotic or necrotic stimuli in different cell types by several mechanisms [50]. Increased expression of HK2 upon LPS stimulation may contribute to pro-survival effects of LPS stimulation in MoDCs. HK-II, but not other HKs, bind and inhibit mTORC1 in the absence of glucose, facilitating autophagy in response to glucose starvation. Thus HK-II can protect cells from cellular damage and provide energy by recycling intracellular constituents [51].

LPS stimulation also increases the expression of the pyruvate kinase PKM2, the last step rate-limiting glycolytic enzyme. In addition to its role in glycolysis, this protein can bind to HIF-1α in the nucleus and regulate the transactivation domain function of HIF-1α, altering the expression of target genes, enhancing PD-L1 expression by BM-DMs and contributing to the production of pro-inflammatory IL-1β. This function is inhibited when PKM2 is stabilized as a tetramer by activators [52, 53].

4.2 TCA Rewiring and Oxidative Phosphorylation

Metabolomics of immune cells recently established that TLR4 stimulation not only induces a glycolytic reprogramming but also the rewiring of TCA cycle that generates inflammatory intermediates contributing to macrophages and DCs activation. LPS stimulation of macrophages leads to transient accumulation of succinate and citrate, highlighting two breakpoints in the TCA cycle [54, 55] (Fig. 2).

Succinate is a proinflammatory metabolite, enhancing LPS-induced HIF-1α activity and driving IL-1β production without affecting TNFα secretion in murine macrophages [25, 26]. Increased succinate oxidation by SDH, which also acts as the complex II of the electron transfer chain, combined with an increased mitochondrial membrane potential are required to generate mitochondrial ROS upon LPS stimulation [26]. This leads to a reverse electron transport from complex II to complex I that produces ROS [56]. Metformin which inhibits complex I, inhibits ROS production, and decreases LPS-induced IL-1β expression [57]. Thus, LPS stimulation of macrophages reorients mitochondrial activity from ATP synthesis to ROS production, favoring HIF-1α activation and activating NLRP3 inflammasome, promoting glycolysis, and enhancing the production of the pro-inflammatory cytokine IL-1β [56]. Succinate can also be secreted from LPS-activated macrophages and activate its cognate receptor SUCNR1 (previously GPR91) in an autocrine and paracrine manner to further enhance production of IL-1β [58]. Succinate-SUCNR1 signaling can act as a chemotactic factor for DCs, enhancing their activation in synergy with TLR signaling [59]. Signaling of succinate through SUCNR1 in myeloid cells has been implicated in exacerbating and sustaining inflammation in chronic pathological conditions including rheumatoid arthritis and obesity [58, 60]. Recent findings have challenged this idea since activation of SUCNR1 can also promote an anti-inflammatory program in macrophages, favoring M2 polarization and suggesting it may play a role in limiting inflammation to regulate the metabolic response to obesity [61].

Reduced IDH expression in M1 pro-inflammatory macrophages leads to citrate accumulation [28]. In addition to fuel fatty acid biosynthesis (cf. above), citrate can be converted into aconitate and then itaconate by the enzyme encoded by the immuno-responsive gene 1 (Irg1), whose expression is enhanced in LPS-stimulated macrophages. Itaconate inhibits SDH activity, making a link between citrate and succinate accumulation. It thus controls mitochondrial respiration changes in M1 macrophages. Itaconate exerts anti-inflammatory effects, reducing LPS-induced production of IL-1β, IL-12, and IL-6 and ROS especially [62, 63]. Although succinate and SDH activity are required to generate proinflammatory response in LPS-stimulated macrophages, sustained SDH activity is likely to have detrimental effects by inducing excessive ROS production [56]. Thus, the regulation of inflammation could be linked to the balanced production of both succinate and itaconate.

LPS stimulation also results in pyruvate dehydrogenase (PDH) inhibition, thereby preventing the generation of acetyl-CoA for the TCA cycle (Fig. 1) in activated murine macrophages [30]. This has been recently characterized as a late response to LPS + IFNγ stimulation of macrophages [55]. Moreover, the expression of several mitochondrial enzymes involved in the TCA cycle are inhibited in LPS-treated macrophages [30].

In BM-DMs and BM-DCs, LPS stimulation results in induction of iNOS expression by several pathways activating HIF-1α described above, resulting in the production of NO from the metabolism of arginine. NO is a diffusible radical that nitrosylates and inhibits complex I, III, and IV of the mitochondrial electron transport chain, reducing O_2 consumption and ATP production. This is likely to explain the impaired mitochondrial OXPHOS observed in M1 murine macrophages and activated BM-DCs. As a consequence, the glycolytic activity of BM-DCs is increased to provide enough ATP for cell survival [14]. Everts et al. have shown that the late, but not early, increase of glycolysis depends on this mechanism [10]. This pathway is not involved in human MoDCs since this inducible NOS is not expressed [64] and NO production triggered by LPS could not be detected in these cells [11].

4.3 Lipid Metabolism

TLR4 stimulation of macrophages results in increased fatty acid uptake and storage as diacyl- or triacyl-glycerol (TAG) in cytosolic lipid droplets [65, 66]. LPS stimulation also decreased lipolysis and FAO, thus favoring TAG accumulation. Several mechanisms converge to increase fatty acid synthesis (FAS) in activated macrophages. Intracellular lipid storage is increased by enhanced long-lasting expression of diacylglycerol acyltransferase-2 (DGAT2) and long-chain acyl-CoA synthetase 1 (ACSL1) [65] or glycerol-3-phosphate acyltransferase 3 (GPAT3) [66]. The incorporation of carbons derived from glucose into FA increases [66], resulting especially from the reorientation of citrate produced by the TCA cycle to FAS [29]. The mitochondrial citrate transporter allowing citrate efflux was found to be essential to ROS, NO, and prostaglandin E2 generation in a LPS-stimulated human macrophage cell line [29]. Citrate metabolism in the cytosol regenerates both acetyl-CoA and oxaloacetate. Acetyl-CoA is used to synthesize FAs that are incorporated into TAG and phospholipids or prostaglandins while oxaloacetate is metabolized into pyruvate, producing NADPH the cofactor for both NADPH oxidase and iNOS [29, 31]. ROS can stabilize HIF-1α and activate NLRP3 inflammasome, promoting glycolysis and enhancing the production of the proinflammatory cytokine IL-1β. Lipid metabolism has also been linked to NLRP3 since reduced FA synthesis decreased NLRP3 activation. Fatty acid synthetase (FASN) regulates NLRP3 and IL-1β expression through AKT activation in LPS-activated murine macrophages [67]. In BM-DCs, enhanced FAS upon TLR4 stimulation required increased glycolysis to generate NADPH through PPP and TCA rewiring to generate citrate [10]. Genetic silencing of mitochondrial citrate transporter or inhibition of acetyl-CoA carboxylase (ACC) inhibiting FAS both impair LPS-induced BM-DC maturation [10].

In mitochondria, FAO catabolizes FA into acetyl-CoA, fueling the TCA cycle and OXPHOS, and providing NADH and FADH. FAO is differentially regulated in M1 and M2 macrophages. M2 macrophages have an increased FAO due to IL-4 stimulation of STAT6 and activation of AMP-activated protein kinase (AMPK) [56]. As an energy sensor, AMPK is activated when AMP is low, inhibiting anabolic pathways and activating catabolic pathways such as FAO. Boosting FAO may thus restrain inflammatory macrophage function. During the immunotolerant phase of sepsis, there is a switch to an increased FAO [68]. Induction of FAO can also be a compensatory mechanism to maintain levels of ATP, NADH, and NADPH in monocytes, when glucose is unavailable, thus preventing the Warburg shift of their metabolism [69]. Conversely, LPS inhibits AMPK resulting in decreased β-oxidation of FA in M1 macrophages. Macrophages and DCs derived from AMPK-α1$^{-/-}$ mice present an enhanced response to TLR4 stimulation, secreting more IL-6 and TNFα inflammatory cytokines [70]. In human MoDCs, AMPK downregulation by LPS also results in increased CD86 co-stimulatory molecule expression and increased IL-12 secretion [9].

5 Conclusion

Numerous studies now indicate that the function of innate immune cells is closely related to their metabolic status. TLR4 stimulation in macrophages or DCs triggers intracellular events, leading to metabolic reprogramming that is essential for their activation and functional maturation. Moreover, sensors of metabolic status can interfere with the response to TLR4 in particular and PRRs in general. Fine tuning of cell metabolism by TLR4 signaling differs according to cell types, species and origin of the cells. The metabolic status of the cells also impacts trained immunity through various metabolites involved in epigenetic regulations. We have reviewed the diversity of mechanisms controlling the metabolic reprogramming of innate immune cells stimulated by TLR4, focusing on glucose and lipid metabolism. Most of the work has been performed in mouse cells and extensive studies exploring the molecular mechanisms involved in the regulation of glycolysis in primary human DCs are warranted for a better understanding of the reciprocal interactions between their cellular metabolic activity and functional activation. The immunometabolism field is emerging and the complex intertwined relationship between immunity and metabolism regulations will likely provide unexpected therapeutic opportunities.

References

1. Mathis D, Shoelson SE. Immunometabolism: an emerging frontier. Nat Rev Immunol. 2011;11(2):81. https://doi.org/10.1038/nri2922.
2. Cohn ZA, Morse SI. Functional and metabolic properties of polymorphonuclear leucocytes. II. The influence of a lipopolysaccharide endotoxin. J Exp Med. 1960;111:689–704. https://doi.org/10.1084/jem.111.5.689.
3. Sbarra AJ, Shirley W. Phagocytosis inhibition and reversal. I. Effect of glycolytic intermediates and nucleotides on particle uptake. J Bacteriol. 1963;86:259–65.
4. Loftus RM, Finlay DK. Immunometabolism: Cellular metabolism turns immune regulator. J Biol Chem. 2016;291(1):1–10. https://doi.org/10.1074/jbc.R115.693903.
5. O'Neill LA, Pearce EJ. Immunometabolism governs dendritic cell and macrophage function. J Exp Med. 2016;213(1):15–23. https://doi.org/10.1084/jem.20151570.
6. Perrin-Cocon L, Aublin-Gex A, Diaz O, Ramiere C, Peri F, Andre P, Lotteau V. Toll-like receptor 4-Induced glycolytic burst in human monocyte-derived dendritic cells results from p38-dependent stabilization of HIF-1alpha and increased hexokinase II expression. J Immunol. 2018;201(5):1510–21. https://doi.org/10.4049/jimmunol.1701522.
7. Jantsch J, Chakravortty D, Turza N, Prechtel AT, Buchholz B, Gerlach RG, Volke M, Glasner J, Warnecke C, Wiesener MS, Eckardt KU, Steinkasserer A, Hensel M, Willam C. Hypoxia and hypoxia-inducible factor-1 alpha modulate lipopolysaccharide-induced dendritic cell activation and function. J Immunol. 2008;180(7):4697–705.
8. Warburg O. On the origin of cancer cells. Science. 1956;123(3191):309–14.
9. Krawczyk CM, Holowka T, Sun J, Blagih J, Amiel E, DeBerardinis RJ, Cross JR, Jung E, Thompson CB, Jones RG, Pearce EJ. Toll-like receptor-induced changes in glycolytic metabolism regulate dendritic cell activation. Blood. 2010;115(23):4742–9. https://doi.org/10.1182/blood-2009-10-249,540.

10. Everts B, Amiel E, Huang SC, Smith AM, Chang CH, Lam WY, Redmann V, Freitas TC, Blagih J, van der Windt GJ, Artyomov MN, Jones RG, Pearce EL, Pearce EJ. TLR-driven early glycolytic reprogramming via the kinases TBK1-IKKvarepsilon supports the anabolic demands of dendritic cell activation. Nat Immunol. 2014;15(4):323–32. https://doi.org/10.1038/ni.2833.
11. Perrin-Cocon L, Aublin-Gex A, Sestito SE, Shirey KA, Patel MC, Andre P, Blanco JC, Vogel SN, Peri F, Lotteau V. TLR4 antagonist FP7 inhibits LPS-induced cytokine production and glycolytic reprogramming in dendritic cells, and protects mice from lethal influenza infection. Sci Rep. 2017;7:40791. https://doi.org/10.1038/srep40791.
12. Guak H, Al Habyan S, Ma EH, Aldossary H, Al-Masri M, Won SY, Ying T, Fixman ED, Jones RG, McCaffrey LM, Krawczyk CM. Glycolytic metabolism is essential for CCR7 oligomerization and dendritic cell migration. Nat Commun. 2018;9(1):2463. https://doi.org/10.1038/s41467-018-04804-6.
13. Thwe PM, Pelgrom L, Cooper R, Beauchamp S, Reisz JA, D'Alessandro A, Everts B, Amiel E. Cell-intrinsic glycogen metabolism supports early glycolytic reprogramming required for dendritic cell immune responses. Cell Metab. 2017;26(3):558–567 e555. https://doi.org/10.1016/j.cmet.2017.08.012.
14. Everts B, Amiel E, van der Windt GJ, Freitas TC, Chott R, Yarasheski KE, Pearce EL, Pearce EJ. Commitment to glycolysis sustains survival of NO-producing inflammatory dendritic cells. Blood. 2012;120(7):1422–31. https://doi.org/10.1182/blood-2012-03-419,747.
15. Cortese M, Sinclair C, Pulendran B. Translating glycolytic metabolism to innate immunity in dendritic cells. Cell Metab. 2014;19(5):737–9. https://doi.org/10.1016/j.cmet.2014.04.012.
16. Vijayan V, Pradhan P, Braud L, Fuchs HR, Gueler F, Motterlini R, Foresti R, Immenschuh S. Human and murine macrophages exhibit differential metabolic responses to lipopolysaccharide – a divergent role for glycolysis. Redox Biol. 2019;22:101147. https://doi.org/10.1016/j.redox.2019.101147.
17. Schmidt EA, Fee BE, Henry SC, Nichols AG, Shinohara ML, Rathmell JC, MacIver NJ, Coers J, Ilkayeva OR, Koves TR, Taylor GA. Metabolic Alterations Contribute to Enhanced Inflammatory Cytokine Production in Irgm1-deficient Macrophages. J Biol Chem. 2017;292(11):4651–62. https://doi.org/10.1074/jbc.M116.770735.
18. Artyomov MN, Sergushichev A, Schilling JD. Integrating immunometabolism and macrophage diversity. Semin Immunol. 2016;28(5):417–24. https://doi.org/10.1016/j.smim.2016.10.004.
19. Lavrich KS, Speen AM, Ghio AJ, Bromberg PA, Samet JM, Alexis NE. Macrophages from the upper and lower human respiratory tract are metabolically distinct. Am J Physiol Lung Cell Mol Physiol. 2018;315(5):L752–64. https://doi.org/10.1152/ajplung.00208.2018.
20. Tannahill GM, O'Neill LA. The emerging role of metabolic regulation in the functioning of Toll-like receptors and the NOD-like receptor Nlrp3. FEBS Lett. 2011;585(11):1568–72. https://doi.org/10.1016/j.febslet.2011.05.008.
21. Malinarich F, Duan K, Hamid RA, Bijin A, Lin WX, Poidinger M, Fairhurst AM, Connolly JE. High mitochondrial respiration and glycolytic capacity represent a metabolic phenotype of human tolerogenic dendritic cells. J Immunol. 2015;194(11):5174–86. https://doi.org/10.4049/jimmunol.1303316.
22. Nagy C, Haschemi A. Time and demand are two critical dimensions of immunometabolism: the process of macrophage activation and the pentose phosphate pathway. Front Immunol. 2015;6:164. https://doi.org/10.3389/fimmu.2015.00164.
23. Haschemi A, Kosma P, Gille L, Evans CR, Burant CF, Starkl P, Knapp B, Haas R, Schmid JA, Jandl C, Amir S, Lubec G, Park J, Esterbauer H, Bilban M, Brizuela L, Pospisilik JA, Otterbein LE, Wagner O. The sedoheptulose kinase CARKL directs macrophage polarization through control of glucose metabolism. Cell Metab. 2012;15(6):813–26. https://doi.org/10.1016/j.cmet.2012.04.023.
24. Bantug GR, Fischer M, Grahlert J, Balmer ML, Unterstab G, Develioglu L, Steiner R, Zhang L, Costa ASH, Gubser PM, Burgener AV, Sauder U, Loliger J, Belle R, Dimeloe S, Lotscher J, Jauch A, Recher M, Honger G, Hall MN, Romero P, Frezza C, Hess C. Mitochondria-endoplasmic reticulum contact sites function as immunometabolic hubs that orchestrate the

rapid recall response of memory CD8(+) T cells. Immunity. 2018;48(3):542–555 e546. https://doi.org/10.1016/j.immuni.2018.02.012.

25. Tannahill GM, Curtis AM, Adamik J, Palsson-McDermott EM, McGettrick AF, Goel G, Frezza C, Bernard NJ, Kelly B, Foley NH, Zheng L, Gardet A, Tong Z, Jany SS, Corr SC, Haneklaus M, Caffrey BE, Pierce K, Walmsley S, Beasley FC, Cummins E, Nizet V, Whyte M, Taylor CT, Lin H, Masters SL, Gottlieb E, Kelly VP, Clish C, Auron PE, Xavier RJ, O'Neill LA. Succinate is an inflammatory signal that induces IL-1beta through HIF-1alpha. Nature. 2013;496(7444):238–42. https://doi.org/10.1038/nature11986.
26. Mills EL, Kelly B, Logan A, Costa ASH, Varma M, Bryant CE, Tourlomousis P, Dabritz JHM, Gottlieb E, Latorre I, Corr SC, McManus G, Ryan D, Jacobs HT, Szibor M, Xavier RJ, Braun T, Frezza C, Murphy MP, O'Neill LA. Succinate dehydrogenase supports metabolic repurposing of mitochondria to drive inflammatory macrophages. Cell. 2016;167(2):457–470 e413. https://doi.org/10.1016/j.cell.2016.08.064.
27. Corcoran SE, O'Neill LA. HIF1alpha and metabolic reprogramming in inflammation. J Clin Invest. 2016;126(10):3699–707. https://doi.org/10.1172/JCI84431.
28. Jha AK, Huang SC, Sergushichev A, Lampropoulou V, Ivanova Y, Loginicheva E, Chmielewski K, Stewart KM, Ashall J, Everts B, Pearce EJ, Driggers EM, Artyomov MN. Network integration of parallel metabolic and transcriptional data reveals metabolic modules that regulate macrophage polarization. Immunity. 2015;42(3):419–30. https://doi.org/10.1016/j.immuni.2015.02.005.
29. Infantino V, Convertini P, Cucci L, Panaro MA, Di Noia MA, Calvello R, Palmieri F, Iacobazzi V. The mitochondrial citrate carrier: a new player in inflammation. Biochem J. 2011;438(3):433–6. https://doi.org/10.1042/BJ20111275.
30. O'Neill LA. A critical role for citrate metabolism in LPS signaling. Biochem J. 2011;438(3):e5–6. https://doi.org/10.1042/BJ20111386.
31. Kelly B, O'Neill LA. Metabolic reprogramming in macrophages and dendritic cells in innate immunity. Cell Res. 2015;25(7):771–84. https://doi.org/10.1038/cr.2015.68.
32. Ibrahim J, Nguyen AH, Rehman A, Ochi A, Jamal M, Graffeo CS, Henning JR, Zambirinis CP, Fallon NC, Barilla R, Badar S, Mitchell A, Rao RS, Acehan D, Frey AB, Miller G. Dendritic cell populations with different concentrations of lipid regulate tolerance and immunity in mouse and human liver. Gastroenterology. 2012;143(4):1061–72. https://doi.org/10.1053/j.gastro.2012.06.003.
33. Netea MG, Joosten LA, Latz E, Mills KH, Natoli G, Stunnenberg HG, O'Neill LA, Xavier RJ. Trained immunity: a program of innate immune memory in health and disease. Science. 2016;352(6284):aaf1098. https://doi.org/10.1126/science.aaf1098.
34. Saeed S, Quintin J, Kerstens HH, Rao NA, Aghajanirefah A, Matarese F, Cheng SC, Ratter J, Berentsen K, van der Ent MA, Sharifi N, Janssen-Megens EM, Ter Huurne M, Mandoli A, van Schaik T, Ng A, Burden F, Downes K, Frontini M, Kumar V, Giamarellos-Bourboulis EJ, Ouwehand WH, van der Meer JW, Joosten LA, Wijmenga C, Martens JH, Xavier RJ, Logie C, Netea MG, Stunnenberg HG. Epigenetic programming of monocyte-to-macrophage differentiation and trained innate immunity. Science. 2014;345(6204):1251086. https://doi.org/10.1126/science.1251086.
35. Foster SL, Hargreaves DC, Medzhitov R. Gene-specific control of inflammation by TLR-induced chromatin modifications. Nature. 2007;447(7147):972–8. https://doi.org/10.1038/nature05836.
36. Yoshida K, Maekawa T, Zhu Y, Renard-Guillet C, Chatton B, Inoue K, Uchiyama T, Ishibashi K, Yamada T, Ohno N, Shirahige K, Okada-Hatakeyama M, Ishii S. The transcription factor ATF7 mediates lipopolysaccharide-induced epigenetic changes in macrophages involved in innate immunological memory. Nat Immunol. 2015;16(10):1034–43. https://doi.org/10.1038/ni.3257.
37. van der Meer JW, Joosten LA, Riksen N, Netea MG. Trained immunity: a smart way to enhance innate immune defence. Mol Immunol. 2015;68(1):40–4. https://doi.org/10.1016/j.molimm.2015.06.019.

38. Cheng SC, Quintin J, Cramer RA, Shepardson KM, Saeed S, Kumar V, Giamarellos-Bourboulis EJ, Martens JH, Rao NA, Aghajanirefah A, Manjeri GR, Li Y, Ifrim DC, Arts RJ, van der Veer BM, Deen PM, Logie C, O'Neill LA, Willems P, van de Veerdonk FL, van der Meer JW, Ng A, Joosten LA, Wijmenga C, Stunnenberg HG, Xavier RJ, Netea MG. mTOR- and HIF-1alpha-mediated aerobic glycolysis as metabolic basis for trained immunity. Science. 2014;345(6204):1250684. https://doi.org/10.1126/science.1250684.
39. Liu TF, Vachharajani VT, Yoza BK, McCall CE. NAD+-dependent Sirtuin 1 and 6 proteins coordinate a switch from glucose to fatty acid oxidation during the acute inflammatory response. J Biol Chem. 2012;287(31):25758–25,769. https://doi.org/10.1074/jbc.M112.362343.
40. Arts RJ, Novakovic B, Ter Horst R, Carvalho A, Bekkering S, Lachmandas E, Rodrigues F, Silvestre R, Cheng SC, Wang SY, Habibi E, Goncalves LG, Mesquita I, Cunha C, van Laarhoven A, van de Veerdonk FL, Williams DL, van der Meer JW, Logie C, O'Neill LA, Dinarello CA, Riksen NP, van Crevel R, Clish C, Notebaart RA, Joosten LA, Stunnenberg HG, Xavier RJ, Netea MG. Glutaminolysis and fumarate accumulation integrate immunometabolic and epigenetic programs in trained immunity. Cell Metab. 2016;24(6):807–19. https://doi.org/10.1016/j.cmet.2016.10.008.
41. Benit P, Letouze E, Rak M, Aubry L, Burnichon N, Favier J, Gimenez-Roqueplo AP, Rustin P. Unsuspected task for an old team: succinate, fumarate and other Krebs cycle acids in metabolic remodeling. Biochim Biophys Acta. 2014;1837(8):1330–7. https://doi.org/10.1016/j.bbabio.2014.03.013.
42. Siegert I, Schodel J, Nairz M, Schatz V, Dettmer K, Dick C, Kalucka J, Franke K, Ehrenschwender M, Schley G, Beneke A, Sutter J, Moll M, Hellerbrand C, Wielockx B, Katschinski DM, Lang R, Galy B, Hentze MW, Koivunen P, Oefner PJ, Bogdan C, Weiss G, Willam C, Jantsch J. Ferritin-mediated Iron sequestration stabilizes hypoxia-inducible factor-1alpha upon LPS activation in the presence of ample oxygen. Cell Rep. 2015;13(10):2048–55. https://doi.org/10.1016/j.celrep.2015.11.005.
43. Spirig R, Djafarzadeh S, Regueira T, Shaw SG, von Garnier C, Takala J, Jakob SM, Rieben R, Lepper PM. Effects of TLR agonists on the hypoxia-regulated transcription factor HIF-1alpha and dendritic cell maturation under normoxic conditions. PLoS ONE. 2010;5(6):e0010983. https://doi.org/10.1371/journal.pone.0010983.
44. Schito L, Semenza GL. Hypoxia-inducible factors: master regulators of cancer progression. Trends Cancer. 2016;2(12):758–70. https://doi.org/10.1016/j.trecan.2016.10.016.
45. Lawless SJ, Kedia-Mehta N, Walls JF, McGarrigle R, Convery O, Sinclair LV, Navarro MN, Murray J, Finlay DK. Glucose represses dendritic cell-induced T cell responses. Nat Commun. 2017;8:15620. https://doi.org/10.1038/ncomms15620.
46. Snyder JP, Amiel E. Regulation of dendritic cell immune function and metabolism by cellular nutrient sensor mammalian Target of Rapamycin (mTOR). Front Immunol. 2018;9:3145. https://doi.org/10.3389/fimmu.2018.03145.
47. Frede S, Stockmann C, Freitag P, Fandrey J. Bacterial lipopolysaccharide induces HIF-1 activation in human monocytes via p44/42 MAPK and NF-kappaB. Biochem J. 2006;396(3):517–27. https://doi.org/10.1042/BJ20051839.
48. Rius J, Guma M, Schachtrup C, Akassoglou K, Zinkernagel AS, Nizet V, Johnson RS, Haddad GG, Karin M. NF-kappaB links innate immunity to the hypoxic response through transcriptional regulation of HIF-1alpha. Nature. 2008;453(7196):807–11. https://doi.org/10.1038/nature06905.
49. Nicholas SA, Sumbayev VV. The role of redox-dependent mechanisms in the downregulation of ligand-induced toll-like receptors 7, 8 and 4-mediated HIF-1 alpha prolyl hydroxylation. Immunol Cell Biol. 2010;88(2):180–6. https://doi.org/10.1038/icb.2009.76.
50. Roberts DJ, Miyamoto S. Hexokinase II integrates energy metabolism and cellular protection: akting on mitochondria and TORCing to autophagy. Cell Death Differ. 2015;22(2):248–57. https://doi.org/10.1038/cdd.2014.173.

51. Roberts DJ, Tan-Sah VP, Ding EY, Smith JM, Miyamoto S. Hexokinase-II positively regulates glucose starvation-induced autophagy through TORC1 inhibition. Mol Cell. 2014;53(4):521–33. https://doi.org/10.1016/j.molcel.2013.12.019.
52. Palsson-McDermott EM, Curtis AM, Goel G, Lauterbach MA, Sheedy FJ, Gleeson LE, van den Bosch MW, Quinn SR, Domingo-Fernandez R, Johnston DG, Jiang JK, Israelsen WJ, Keane J, Thomas C, Clish C, Vander Heiden M, Xavier RJ, O'Neill LA. Pyruvate kinase M2 regulates Hif-1alpha activity and IL-1beta induction and is a critical determinant of the warburg effect in LPS-activated macrophages. Cell Metab. 2015;21(1):65–80. https://doi.org/10.1016/j.cmet.2014.12.005.
53. Palsson-McDermott EM, Dyck L, Zaslona Z, Menon D, McGettrick AF, Mills KHG, O'Neill LA. Pyruvate kinase M2 is required for the expression of the immune checkpoint PD-L1 in immune cells and tumors. Front Immunol. 2017;8:1300. https://doi.org/10.3389/fimmu.2017.01300.
54. O'Neill LA. A broken Krebs cycle in macrophages. Immunity. 2015;42(3):393–4. https://doi.org/10.1016/j.immuni.2015.02.017.
55. Seim GL, Britt EC, John SV, Yeo FJ, Johnson AR, Eisenstein RS, Pagliarini DJ, Fan J. Two-stage metabolic remodelling in macrophages in response to lipopolysaccharide and interferon-γ stimulation. Nat Metab. 2019;1(7):731–42. https://doi.org/10.1038/s42255-019-0083-2.
56. Mills EL, Kelly B, O'Neill LAJ. Mitochondria are the powerhouses of immunity. Nat Immunol. 2017;18(5):488–98. https://doi.org/10.1038/ni.3704.
57. Kelly B, Tannahill GM, Murphy MP, O'Neill LA. Metformin inhibits the production of reactive oxygen species from NADH: ubiquinone oxidoreductase to limit induction of interleukin-1beta (IL-1beta) and boosts interleukin-10 (IL-10) in lipopolysaccharide (LPS)-activated macrophages. J Biol Chem. 2015;290(33):20348–20,359. https://doi.org/10.1074/jbc.M115.662114.
58. Littlewood-Evans A, Sarret S, Apfel V, Loesle P, Dawson J, Zhang J, Muller A, Tigani B, Kneuer R, Patel S, Valeaux S, Gommermann N, Rubic-Schneider T, Junt T, Carballido JM. GPR91 senses extracellular succinate released from inflammatory macrophages and exacerbates rheumatoid arthritis. J Exp Med. 2016;213(9):1655–62. https://doi.org/10.1084/jem.20160061.
59. Rubic T, Lametschwandtner G, Jost S, Hinteregger S, Kund J, Carballido-Perrig N, Schwarzler C, Junt T, Voshol H, Meingassner JG, Mao X, Werner G, Rot A, Carballido JM. Triggering the succinate receptor GPR91 on dendritic cells enhances immunity. Nat Immunol. 2008;9(11):1261–9. https://doi.org/10.1038/ni.1657.
60. van Diepen JA, Robben JH, Hooiveld GJ, Carmone C, Alsady M, Boutens L, Bekkenkamp-Grovenstein M, Hijmans A, Engelke UFH, Wevers RA, Netea MG, Tack CJ, Stienstra R, Deen PMT. SUCNR1-mediated chemotaxis of macrophages aggravates obesity-induced inflammation and diabetes. Diabetologia. 2017;60(7):1304–13. https://doi.org/10.1007/s00125-017-4261-z.
61. Keiran N, Ceperuelo-Mallafre V, Calvo E, Hernandez-Alvarez MI, Ejarque M, Nunez-Roa C, Horrillo D, Maymo-Masip E, Rodriguez MM, Fradera R, de la Rosa JV, Jorba R, Megia A, Zorzano A, Medina-Gomez G, Serena C, Castrillo A, Vendrell J, Fernandez-Veledo S. SUCNR1 controls an anti-inflammatory program in macrophages to regulate the metabolic response to obesity. Nat Immunol. 2019;20(5):581–92. https://doi.org/10.1038/s41590-019-0372-7.
62. Cordes T, Wallace M, Michelucci A, Divakaruni AS, Sapcariu SC, Sousa C, Koseki H, Cabrales P, Murphy AN, Hiller K, Metallo CM. Immunoresponsive gene 1 and itaconate inhibit succinate dehydrogenase to modulate intracellular succinate levels. J Biol Chem. 2016;291(27):14274–14,284. https://doi.org/10.1074/jbc.M115.685792.
63. Lampropoulou V, Sergushichev A, Bambouskova M, Nair S, Vincent EE, Loginicheva E, Cervantes-Barragan L, Ma X, Huang SC, Griss T, Weinheimer CJ, Khader S, Randolph GJ, Pearce EJ, Jones RG, Diwan A, Diamond MS, Artyomov MN. Itaconate links inhibition of

succinate dehydrogenase with macrophage metabolic remodeling and regulation of inflammation. Cell Metab. 2016;24(1):158–66. https://doi.org/10.1016/j.cmet.2016.06.004.
64. Schneemann M, Schoeden G. Macrophage biology and immunology: man is not a mouse. J Leukoc Biol. 2007;81(3):579–80. https://doi.org/10.1189/jlb.1106702.
65. Huang YL, Morales-Rosado J, Ray J, Myers TG, Kho T, Lu M, Munford RS. Toll-like receptor agonists promote prolonged triglyceride storage in macrophages. J Biol Chem. 2014;289(5):3001–12. https://doi.org/10.1074/jbc.M113.524587.
66. Feingold KR, Shigenaga JK, Kazemi MR, McDonald CM, Patzek SM, Cross AS, Moser A, Grunfeld C. Mechanisms of triglyceride accumulation in activated macrophages. J Leukoc Biol. 2012;92(4):829–39. https://doi.org/10.1189/jlb.1111537.
67. Moon JS, Lee S, Park MA, Siempos II, Haslip M, Lee PJ, Yun M, Kim CK, Howrylak J, Ryter SW, Nakahira K, Choi AM. UCP2-induced fatty acid synthase promotes NLRP3 inflammasome activation during sepsis. J Clin Invest. 2015;125(2):665–80. https://doi.org/10.1172/JCI78253.
68. Liu TF, Vachharajani V, Millet P, Bharadwaj MS, Molina AJ, McCall CE. Sequential actions of SIRT1-RELB-SIRT3 coordinate nuclear-mitochondrial communication during immunometabolic adaptation to acute inflammation and sepsis. J Biol Chem. 2015;290(1):396–408. https://doi.org/10.1074/jbc.M114.566349.
69. Raulien N, Friedrich K, Strobel S, Rubner S, Baumann S, von Bergen M, Korner A, Krueger M, Rossol M, Wagner U. Fatty acid oxidation compensates for lipopolysaccharide-induced warburg effect in glucose-deprived monocytes. Front Immunol. 2017;8:609. https://doi.org/10.3389/fimmu.2017.00609.
70. Carroll KC, Viollet B, Suttles J. AMPKalpha1 deficiency amplifies proinflammatory myeloid APC activity and CD40 signaling. J Leukoc Biol. 2013;94(6):1113–21. https://doi.org/10.1189/jlb.0313157.

Correction to: The Role of Toll-Like Receptor 4 in Infectious and Non Infectious Inflammation

Carlo Rossetti and Francesco Peri

Correction to:
C. Rossetti, F. Peri (eds.), *The Role of Toll-Like Receptor 4 in Infectious and Non Infectious Inflammation*, Progress in Inflammation Research 87, https://doi.org/10.1007/978-3-030-56319-6

This book published under the series "Progress in Inflammation Research" was inadvertently published with the incorrect series ISSN number. This has now been amended in the book to the correct ISSN:

Print ISSN: 1422-7746

Electronic ISSN: 2296-4525

The updated version of the book can be found at https://doi.org/10.1007/978-3-030-56319-6

C. Rossetti, F. Peri (eds.), *The Role of Toll-Like Receptor 4 in Infectious and Non Infectious Inflammation*, Progress in Inflammation Research 87,
https://doi.org/10.1007/978-3-030-56319-6_12

Printed in the United States
by Baker & Taylor Publisher Services